90 0399818 0

3

THE PROBLEMS OF REFUGEES IN AFRICA

Dedicated to my wife, Cynthia and my children, Cyril, Dorothy and Margaret.
Thank you for being so loyal and supportive.

We must transform the presence of refugees in the various African countries, which is often felt to be a liability, into assets of the economic and social balance of the countries' development ...

Prince Sadruddin Aga Khan
former UN High Commissioner for Refugees,
Conference on the Legal, Economic and Social Aspects of African Refugee
Problems, 9–18 October, 1967, Final Report. Sponsors: United Nations Economic
Commission for Africa; United Nations High Commissioner for Refugees;
Organisation of African Unity; Dag Hammarskjold Foundation. Addis Ababa,
December, 1968

The Problems of Refugees in Africa

Boundaries and borders

EBENEZER Q. BLAVO
Centre for Social Policy Studies
University of Ghana
Legon

Ashgate

Aldershot • Brookfield USA • Singapore • Sydney

Published by
Ashgate Publishing Ltd
Gower House
Croft Road
Aldershot
Hants GU11 3HR
England

Ashgate Publishing Company
Old Post Road
Brookfield
Vermont 05036
USA

British Library Cataloguing in Publication Data
Blavo, Ebenezer Quarshie
    The problems of refugees in Africa: boundaries and
    borders. - (University of North London voices in
    development management)
    1. Refugees - Africa 2. Refugees - Government policy - Africa
    I. Title II. University of North London
    325.2'1'096

Library of Congress Catalog Card Number: 99-72846

ISBN 1 84014 999 X ✓

Printed in Great Britain by
Antony Rowe Ltd, Chippenham, Wiltshire

Contents

Acknowledgments

The refugee situation in Africa is the direct outcome of the numerous conflicts plaguing almost all the countries in Africa since they attained independence. It has created tremendous physical, social and psychological problems for the uprooted, has wasted vast material resources and impeded progress in the developmental efforts of governments, Many are unaware of the death-toll, the bitter experiences associated with homelessness and what it means to be in refuge. This book is an attempt to make the facts known to all and to stress the need for a concerted effort by all to achieve lasting peace on the continent.

My service to the United Nations High Commissioner for Refugees 1973–1985 gave me ample opportunity to be in the field and in close contact with refugees of almost all nationalities in Africa, and with operational partners, government agencies and non-governmental organizations, including the religious bodies. I came into grips with the real situation and appreciated the confidences entrusted to me by refugees. This heightened my resolve and the commitment to their cause. Together with government counterparts in many African countries, we promoted our joint endeavours to find solutions to the complex problems on our hands. All this knowledge and experience has been brought to bear on the writing of this book. I am greatly indebted to all of them.

This work was originally in partial fulfilment of the requirements for the award of the Doctor of Philosophy Degree in Sociology, University of Ghana. I was encouraged by friends and former colleagues to publish my work at this time of extensive turmoil in Africa and to highlight current world concerns for refugees.

I would like to acknowledge the assistance of one such friend, Margaret Grieco, DPhil (Oxon), Professor of Organisation and Development Management of the Business School of the University of North London, by whose guidance and support this book was published. I am truly grateful to her.

I am indebted to my wife Cynthia for showing considerable interest in my work.

I express my appreciation for the untiring effort of Mrs Dorothy Stella Quarshie in typing the manuscript. Finally, I would like to thank Ashgate Publishing Limited for publishing this book. Needless to say I take full responsibility for everything written in this book.

Ebenezer Q. Blavo
Affiliate
Centre for Social Policy Studies
University of Ghana
Legon

Acronyms

AACC	All African Council of Churches
ECA	United Nations Economic Commission for Africa
ICRC	International Committee for the Red Cross
ILO	International Labour Organization
LWF	Lutheran World Federation
OAU	Organization of African Unity
UN	United Nations Organization
UNDP	United Nations Development Programme
UNHCR	United Nations Economic Commissioner for Refugees
UNICEF	United Nations Children's Fund
WCC	World Council of Churches
WFP	World Food Programme
WHO	World Health Organization

Introduction

The Problem

Africa is going through a crucial period of adversity this century. In the past five decades, various forms of man-made disasters have forced millions of people out of their homeland into refuge in other countries. The uprooted are essentially women and children exposed to life-threatening situations, the consequences of which they are destined to endure. While in refuge, theirs is a desperate struggle for protection and sustenance.

Uprootment is not new to mankind. From time immemorial individuals and large groups of people had to take refuge long distances away from home in the face of persecution - wars and terror tactics of rulers. The modern phenomenon of uprootment in Africa is the direct outcome of the struggle for independence from colonial rule. The post-independence era is marred by conflicts emanating, in the main, from ethnic upheavals suppressed during colonial rule; border disputes over artificial boundaries created during the partitioning of the continent; confrontations between opposing political parties, and human rights abuses. Almost all of Africa is involved; each country either producing or receiving refugees or both, however insignificant.

The economic consequences of these conflicts directly impede both the social adjustment of refugees in their countries of refuge, and the development of the countries involved. A great deal of material resources are put to waste in conflict while nationals and refugees alike live in want.

The importance of the world refugee problem came to international awareness during World War II.

In peace time in 1951 the United Nations adopted the Convention Relating to the Status of Refugees to promote the interest of refugees in Europe.[1] When the refugee problem in Africa reached unmanageable proportions in the late 1950s, the United Nations promulgated the Protocol Relating to the Status of Refugees (1967) as an extension to the Convention (1951) in order to promote the interest of African refugees.[2] Since then African governments have created the Organization of African Unity (OAU), and under its auspices promulgated the OAU Convention Governing the Specific Aspects of Refugee Problems

1

in Africa (1969).[3] Contracting states to the International and Regional Refugee Conventions have adopted national refugee policies, based on the Conventions, for their guidance in the management of refugees in their territories.

This text focuses attention basically on the problems of refugees in Africa, and proposes the best possible means of finding solutions. An effort has been made to point out the basic causes of the refugee problem in Africa as political issues of international dimension urging governments to resolve conflicts and ensure lasting peace.

The problems of refugees are the physical hardships they endure when seeking refuge and the bitter experiences they encounter in forging a new life in refuge. These problems are solved from the perspective of making meaningful national refugee policies in a spirit of brotherhood so that refugees receive cordial reception and a warm hospitality to counteract their anxieties. Such policies give refugees the assurances of their safety till they are able to return to their homeland.

Structure of the Book

Chapter 1

'Causes of Refugee Problems in Africa' highlights the nature of the conflicts in Africa during the period of transition from colonialism to independence and in the post-independence era.

Chapter 2

'The Concept of Refugee' examines the criteria by which refugee status is granted to the uprooted. It focuses on the various humanitarian issues upon which others who may not legally qualify for refugee status are protected, and explains conditions necessary for relinquishing refugee status. The international and regional refugee Conventions are extensively applied for clarification.

Chapter 3

'Life in Refuge' delves into the experiences of refugees in their efforts of adjustment to life in refuge, and establishes the factors that align the problems to crises in the lives of refugees.

Chapter 4

'Solving the Problems of Refugees in Africa' analyses the forms of protection and material support refugees receive in compliance with their legal rights under the International Refugee Convention, and explores all the possible means of ending the period in refuge.

Chapter 5

'Refugee Policies of African Countries' assesses the value and effectiveness of existing refugee policies in Some African countries, endorses the relevance of refugee policies, and makes proposals for setting out meaningful policies in the best interests of refugees.

Notes

1   The Convention Relating to the Status of Refugees (1951) contains a general definition of refugee. It sets down standards of treatment to be accorded refugees. States that are parties to this Convention undertake contractual obligations in an area where customarily they had none.
2   The general definition of refugee in the Convention Relating to the Status of Refugees (1951) is restricted by the provision that refugee status must be the result of events occurring before 1 January 1951. The Protocol removed the dateline of 1 January 1951 in the Convention and made the substantive provisions of the Convention applicable to new refugee groups.
3   The Convention of the Organization of African Unity (1969) supplements the UN Convention (1951) and the Protocol (1967) with provisions that were created to cover the special problems that have arisen in Africa relating to refugees.

1 Causes of Refugee Problems in Africa

The causes of conflicts in Africa this century can be traced far back into history when Europeans first made contact with Africa (Brooks and El-Ayouty, 1970). The economic, social and political structures of the indigenous people were greatly influenced by the Europeans, much to the disadvantage of the people. After centuries of contact with the developed world, Africa is still undeveloped, the quality of life of the people relatively low, and governments have not enough material resources to meet the needs of the people. Much as these economic factors prove provocative, adverse effects on the traditional political system and neglect of human resources development attest to the discontentments that form the basis of the conflicts in Africa.

Initially, the slave trade between Africa and the outside world caused population loss (Rodney, 1972), destabilized society on account of inter-ethnic wars, and weakened traditional rule. Trading with Europe exposed Africans to luxuries and made them materialistic, while their rich natural resources were exploited. Europe found its contact with Africa lucrative and at the Berlin Conference in 1884 the continent was partitioned and colonized. The national boundaries created arbitrarily divided many ethnic communities into different nationalities, thus crippling the already fragile traditional rule. Furthermore, some national boundaries were not clearly defined and more than a century later became a subject of dispute between adjacent countries.

Colonial administration reigned supreme over traditional rule and further weakened this political institution. In order to consolidate European administration, urban centres with basic social services and the utilities were established. A labour market was established resulting in a massive rural-urban drift and neglect of traditional economic activities. Adverse effects were unemployment, and a change in social values and cultural norms. A disintegration of the extended family system adversely affected the social and economic wellbeing of the people.

In order to appreciate the significance of these developments in relation to the causes of conflicts and the uprootment problem, it is worth delving more into the historical perspectives of the economic, social and political systems of the indigenous people.

The African Social System

African communities were characterized by the extended family system where many blood-related nuclear families lived in association under family heads and shared economic and social life. The women cared for the home and the young, and the men tilled the soil and reared animals. The welfare of the sick, the disabled, the orphan and the widow was the responsibility of all members of the extended family.

Culture and religion were preserved by succeeding generations (ibid.), and these served to identify one ethnic community from another.

Ethnic identity was further strengthened by language and dialects, costume, social values, and in some areas by emblems and facial marks.

The social norms pursued in the upbringing and discipline of children permeated the whole fabric of society. The idea of seniority through age was reflected in the presence of age-grades and age-sets among a great many African communities (Nyerere, 1974). Perhaps this respect for age is most clearly in evidence especially among those communities that emphasized age as an essential principle of their social organization. Among these communities are the Kikuyu, Masai, Nandi, Kipsijis, whose social organizations have been elaborately described in anthropological monographs (Huntingford, 1953).

By tradition, Africans live in ethnic communities under recognized leadership of chiefs and elders. This is the political institution installed by the people to govern their lives. The institution had authority over the land and all its wealth, provided security, inspired a sense of patriotism among its subjects, and gave leadership in times of crisis in defence of the community as a whole. The people paid their allegiance to the chief and were under his jurisdiction. Some of these political institutions developed into powerful kingdoms with a monarch as head of many large ethnic communities.

Social Transformation

Traditional social life began to disintegrate at the inception of the slave trade (Hatch, 1960). African rulers found European goods sufficiently appealing, and considered it desirable to hand over captives they had taken in warfare in exchange for foreign goods, including firearms. War became rampant between the kingdoms and communities for the sole purpose of getting captives for sale. Even within a given community a ruler could exploit his subjects and capture them in exchange for material wealth. The slave trade was the beginning of population loss and loss of human potential for development.

The European political administrative structure was an autonomy with the governor at its head. Its entrenchment and security were guaranteed by the numerous forts and castles spread over Africa. Regional and District Commissioners passed on directives of the governor to the chiefs and people. In this set-up the indigenous people had no political rights; they were not involved in the making of any serious administrative policies. The chiefs collected taxes and enforced labour in the mines and the industries. The demand for labour for construction of harbours, roads, railways and administrative centres, and also for military tasks, led to the introduction of forced labour. The coercive methods used to regulate labour recruitment caused much physical suffering and loss of lives. Quite often, in times of agitation, the chiefs suppressed freedom of expression of their subjects. Since the chiefs had lost administrative powers and could no longer give adequate protection to their subjects, they lost respect, and the stability within the ethnic community was seriously crippled.

Traditional rule has survived to the present but it has very limited political power. Some have become fragmented, others have gone into alliances, and generally there are several ethnic communities still in conflict.

Economic Transformation

Economic policies of ethnic communities revolved around agriculture – land use and rearing of animals. Every family was assured sufficient land to meet its subsistence needs. Land was jealously guarded, not only as the basis for economic growth, but also as an ethnic possession and a unifying force.

Contact with the outside world opened up other economic opportunities. By the fifteenth century, Africans everywhere had arrived at a considerable understanding of the total ecology – of the soil, climate, animals, plants and their relationships. In evidence were the techniques for building homes, making utensils, trapping animals, using medicinal plants, and above all developing systems of agriculture (Rodney, 1972).

Advanced methods of agriculture were employed in terracing, crop rotation, green manuring, mixed farming and regulated swamp-farming. The single most important technological change underlying African agricultural development was the introduction of iron tools, notably the axe and the hoe, replacing wooden and stone tools (ibid.). Other artifacts made by hand were the beginnings of industrialization.

Manufactured products included cotton cloth from Guinea, copper ornaments from Katanga and Zambia, iron implements from Sierra Leone, and dye from Yoruba (Nigeria).

Trading was extensive with Europe and the Arab world. Barter was generally practised when the volume of trade was small. As trade became more complicated, a standard of measuring goods was adopted. Salt, cloth, iron hoes and cowry shells were popular forms of money in Africa. Gold and copper, also means of monetary exchange, were restricted to measuring goods of great value. Exports from Africa included sisal, coffee and cotton from Tanganyika (Tanzania); diamonds, oil and coffee from Angola; and cash crops from Morocco. Also traded in were timber, ivory, gold, bauxite and other minerals needed by the industries in Europe.

European agricultural policies brought problems of landownership (Wallerstein, 1971). The chiefs had to release community lands for cultivation of cash crops owned by settlers, thus diverting attention from subsistence agriculture. For instance, under the stresses of social reorganization and resettlement under the British agricultural scheme, the Sudanese Zande community suffered a certain amount of social disintegration, and a diminution of effective Zande leadership. Generally, agricultural policies reduced African economic self-reliance and initiative.

A decisive and fundamental structural change occurred when a labour market was established and jobs were created within the settler administration and services.

This had two major effects on the people. Jobs attracted salaries and wages which served as reward for individual effort. While people became more committed to their jobs communal obligations were neglected. Initially, urbanization stifled rural subsistence farming and those living in urban centres got accustomed to foreign foods and goods. Secondly, the employer-employee system gained impetus over agriculture and changed people's attitude to work. However, as education was a prerequisite to employment and as the rural-drifters were mainly illiterate, no jobs were offered to them. This state of affairs, coupled with limited job openings, brought about unemployment. As the unemployed population increased, society had to bear the brunt of social deterioration. Illiteracy and unemployment problems are rife today and they form some of the causes of discontent and social unrest.

Urbanization had one other serious adverse consequence. Association with the settlers exposed the indigenous people to foreign social values and life-styles. This culminated in loss of significance of African culture and tradition. Nuclear families now lived independently of one another, and this weakened the close-knit extended family system with its safeguards of social security. The presence of vulnerable persons became evident. The aged, the disabled, invalids, the orphan and the widow needed social support, but this time from government. When these changes occurred, only rudimentary social services patterned after those in the metropolis were introduced, but these could not be substituted for those services that the family could no longer perform (Wallerstein, 1971).

Education was primarily to train people to serve the colonial administration (ibid.). The kind of education established was not related to skills needed for the development of the continent. Under the paternalism of Belgian rule in the Congo (Zaire), for instance, education was in no way designed to prepare the Congolese to take real responsibility either in their jobs or in society at large. At independence there was no Congolese in the administration above the level of chief clerk, none in the army above noncommissioned officer, and nowhere in the country was there a Congolese doctor, lawyer or engineer. In the former Portuguese territories, the literacy rate of Africans remained no more than 1 per cent (Rodney, 1972). Health facilities which were made available, catered primarily for the settlers. Areas where the majority of the indigenous people lived were neglected.

The establishment of educational and health facilities were given top priority over other developments in every African country soon after independence from colonial rule. Even so, after several decades of independence, African countries are still handicapped with inadequate social services. Once in virtual poverty and depreciation of the quality of life, coupled with lack of opportunities for advancement, the people's discontentment heightened against the settler administration.

The Conflicts

There was strong resistance against colonial rule at various times in African history and these consolidated into open conflicts with the colonial governments long before independence. The tensions and conflicts that characterized the struggle for independence also had their origins in the system of administration during the colonial period.

Europe's continued dependence on human and material resources from Africa, even when there was massive opposition to their rule, brought resistance to a peak.

The second world war presented the majority of the European nations with possession in Africa, particularly Britain and France, with the most serious threat to their position as colonial powers. The human and material resources of these powers were strained to the utmost, and it became vitally necessary to enlist the physical and moral support of their African colonies which were in a position to provide additional men and certain essential raw materials for the war effort (Sender and Smith, 1986).

The colonial powers loudly proclaimed their intention of giving their subjects greater freedom and better economic opportunities once the war had been brought to a successful conclusion. The Atlantic Charter of 1942, for instance, proclaimed 'the right of all peoples to choose the form of government under which they live'. No sooner had the war ended than the African people began to demand the fulfilment of these promises. Their demands were reinforced by the heavy sacrifices colonial servicemen had made during the war, and their broadened outlook from overseas experience. They demanded equality, not only in the political arena, but also in all aspects of life – economic and social. Their demands eventually developed into well organized and vocal nationalist movements (ibid.), the source of conflict.

From the mid-1950s to the present, almost all African countries have gained independence, but each had to fight hard. Algeria waged war with France. Spanish Sahara (Western Sahara) waged war with Spain. Mozambique and Angola were engaged in a bloody guerilla warfare against the Portuguese. In West Africa, the struggle for independence took the form of internal political strife of varying dimensions; from protests and strike action to harassment of settlers. Many patriotic citizens lost their lives, some were maimed, and others had to flee their country. Some assembled in exile to fight back and many perished in guerilla warfare.

Post-independence conflicts are, in the main, based within the nations and involve the nationals themselves. Nevertheless, these conflicts also were the outcome of the political situation during colonial administration. Discontent with the system of administration of ruling political parties and ethnic confrontations still persist. Constitutionally-elected civilian governments in many African countries have been ousted a number of times by either an opposing civilian political faction or by the military. There have been civil wars between ethnic communities in Nigeria, the Congo, and in Ethiopia and Liberia. Ethnic communities in Rwanda and Burundi have had several bloody confrontations. The Sudan and Chad have been severely destabilized by internal strife. Algeria and Morocco have fought pitched battles over their ill-defined border; Kenya and Somalia have only recently agreed to stop fighting over an ill-defined and, to the local inhabitants, a nonexistent border. The Ethiopia-Somali border at the Ogaden is another area of continual strife.

During Uganda's years as a British protectorate (from mid-1890s until 1962), the traditional kingdoms of Buganda, Bunyoro, Toro and Ankole retained a high degree of autonomy and, in the case of Buganda, the kingdom's social, economic and political dominance was greatly strengthened by British rule.

The new independent nation began as a loose federation that included the four kingdoms. The post-independence forces for disintegration were so strong that success of the central government in holding the country together appeared little short of a miracle.

During the period from 1889 to 1923 when Zambia was administered by the British South African Company, the European population never exceeded 4,000 persons. But with the discovery of copper in 1924, Europeans from South Africa and Southern Rhodesia (Zimbabwe) began arriving in significant numbers. The discovery of copper and the arrival of European settlers had considerable effect on political developments in Zambia. This led to the creation of the Federation of Rhodesia and Nyasaland in 1953 which joined together the two Rhodesias and present-day Malawi. The Federation, largely opposed by the indigenous people in all the three territories, broke up after 10 eventful years (Rodney, 1972).

For 400 years the Hutu majority who made up 85 per cent of the population in Rwanda were dominated by the Tutsi minority, tall slender people who came from the north and established themselves as an aristocracy over the smaller, humbler Hutu people. This feudal structure and its institutions consolidated social, political and economic power almost entirely in the hands of the Tutsi. The feudal system remained intact under German rule from 1910 1916 and under Belgian mandate after World War I. The Belgians for a number of years did little to interfere with Tutsi domination.

Yet in the late 1950s when ethnic-based political parties abruptly sprang into being, it became Belgium's policy to back the newly-emerging Hutu elite and majority rule. This led to the struggle between the parties resulting in a massive refugee problem.

Historically, there had always been a division between the Arab north and the southerners of Sudan who are African in language, culture, and outlook. With the pre-independence replacement of British administrators by the

Sudanese, the majority of whom were northerners, tension between the north and south flared up into open violence resulting in a large number of uprooted people from southern Sudan into neighbouring countries.

It is obvious that separate development of ethnic communities under colonial rule has been one of the causes of political instability in Africa. The political tensions in Chad also originated from separate development by the colonial rulers of the Arab Muslims of the east and north of the country and the Saras and other non-Muslims in the south.

There were other countries in Africa where the idea of African self-rule seemed totally unacceptable to the foreign minority settlers. These were South Africa and Rhodesia (now independent Zimbabwe) where European settlers moved in even before the partitioning of the continent began. There was also Namibia, a United Nations Mandated Territory, which until 23 April 1996, was under South African government control. These governments established oppressive regimes based on discrimination against the indigenous African populations, and their exclusion from any form of participation in serious political affairs. Here it was not the imposition of foreign governments based in Europe that the fight for African freedom had to contend with, but rather the determined resistance of strongly entrenched alien regimes operating within the actual boundaries of the countries concerned. The dislodgement of these regimes proved difficult. The indigenous people of Zimbabwe and Namibia waged a bloody guerrilla war against the settlers to take over political administration. South Africa has only in the past few years achieved success in overcoming the apartheid system and attaining majority rule; but this involved much agitation and bloodshed.

Within the past five decades of turmoil in Africa, much of Africa's resources have gone into conflict and the continent is still lagging behind in development. More importantly, millions of Africans have become victims of circumstances and are in refuge in other African countries. Their anxieties and insecurity are the subject for consideration by host governments and the international community.

2 The Concept of Refugee

The identity of an individual by nationality is an important issue in modern times on account of the responsibilities a state has for its nationals. A national has a right to protection by the government of his country while at home or in a foreign country, In view of this, an immigrant arriving at an international border is under obligation to declare his identity before being allowed entry into the country. This procedure is primarily to ascertain which country is responsible for his welfare.

A person forced out of his home by some tragedy in search of a safe haven may find one within the boundaries of his country, and continue to benefit from the protection and assistance of his government. The term 'internally displaced person' will be used to identify such a person. On the contrary, a refugee is a person who flees into a neighbouring country for refuge in the face of persecution or a major conflict such as civil war, Once inside another sovereign state, he seeks the protection of the government of the country in which he has sought refuge, and no longer has the protection of his own government. His needs for protection and material assistance rest on the host government and on the international community. While in this new environment, the issue of his identity arises, and this includes his race, nationality, sometimes his ethnic origins, and his economic, social and political background.

African refugees represent a fair cross-section of society. They may be individuals or small groups of adults of varying social and political backgrounds such as politicians, members of the armed forces or police, the learned professions, businessmen, civil servants, students and others from urban centres. A typical large rural exodus moving across an international border is composed predominantly of adult females and children.

Among this group are the elderly, the disabled, orphans and unaccompanied minors. Cultural and religious differences abound where there are groups from different countries.

Legal Concept

The issue of identity of the refugee has been considered at international level in much more detail in respect of the suffering masses who fled their homelands during the two world wars.

The earlier International Instruments adopted in the 1920s and 1930s applied the concept of 'Refugee' to specifically enumerated categories of persons who needed international assistance. After World War II, a new international definition was adopted by the United Nations in an Instrument – the 1951 United Nations Convention Relating to the Status of Refugees. In Article I Section A of the Convention, 'Refugee' is a person who:

> As a result of events occurring before 1st January, 1951 and owing to well-founded fear of being persecuted for reasons of race, religion, nationality, membership of a particular social group or political opinion, is outside the country of his nationality and is unable or, owing to such fear, is unwilling to avail himself of the protection of that country; or who not having a nationality and being outside the country of his former habitual residence as a result of such events, is unable or, owing to such fear, is unwilling to return to it (United Nations, Geneva, 28 July 1951).

This definition is of special significance to all contracting or signatory governments in their efforts to identify persons as refugees and to promote their welfare.

The statute adopted by the UN General Assembly setting up the office of the High Commissioner for Refugees in 1950, defines persons of concern to the High Commissioner. This definition is identical to that provided by the International Refugee Convention. As new refugee matters came into focus in various regions of the world, the need was felt by organizations handling refugee problems to make the provisions of the Convention applicable to such new refugees.

The Protocol Relating to the Status of Refugees (United Nations, New York, 31 January 1967) was prepared and submitted to the United Nations General Assembly in 1966 and was adopted on 4 October 1967. The purpose of the Protocol was to expand the International Convention 1951 to embrace all refugees in places other than in Europe, thus eliminating the temporal and geographic limitations in the scope of the Convention.

Although an extension to the UN Refugee Convention 1951, the Protocol is an independent instrument, accession to which is not limited to States that are party to the 1951 Convention. As at July 1987, 103 countries were party to one or both of these instruments (see Appendices), out of which 46 were African countries (see Appendices).

The late 1950s marked the beginning of independence of African countries from colonial rule, albeit, with lots of turmoil and bloodshed. The massive rural exoduses of uprooted persons gave cause for concern. The OAU adopted the Convention Governing the Specific Aspects of Refugee Problems in Africa in 1969 to focus attention on the African refugee situation. The basic elements of the definition of 'refugee' in the UN Refugee Convention were adopted with an extension specifying the circumstances under which African refugees are uprooted. Paragraph 2 of Art. (1) of the OAU Refugee Convention defines 'refugee' as a person who:

> owing to external aggression, occupation, foreign domination or events seriously disturbing public order in either part or the whole of his country of origin or nationality, is compelled to leave his place of habitual residence in order to seek refuge in another place outside his country of origin or nationality.

The operative words 'nationality' and 'persecution' in the International and Regional Refugee Conventions have some important undertones. 'Nationality' implies the country or countries of which a person can prove citizenship by means of documents; or of which he is simply a national by birth or parentage; or where he belongs to a particular ethnic group, and can be identified as such by his culture. Thus, if a person has no such proof, nationality can be taken as membership of a place which he proves to have lived for most of his life, or has habitually lived. This is often referred to as his country of origin. The significance of nationality is the need to relate the person to the circumstances of his uprootment. For instance, if he is not a national of a country in conflict, then it would seem proper for him to go to his own country if he is afraid to stay in that country. Then the question of refuge does not arise.

The person seeking refuge must have 'a well-founded fear of persecution'. Persecution then is a life-threatening situation directed against him personally, or against a group to which he belongs. His well-founded fear can then be established. More importantly, persecution should be seen as having been caused by the government of the country of origin. Or else, perpetrated by a sector of the community in collaboration or connivance with the government. Persecution applies also if it is knowingly tolerated by the government, or when the government, on becoming aware of it, could not offer effective protection. Good reasons must be given for believing that the government is somewhat involved or is connected in some way to the persecution he is suffering or is afraid of. In this context, therefore, a person who is escaping direct or indirect threat from another person or persons unconnected with government is legally not a refugee.

Also, a person who has escaped punishment for a crime he has committed in his country of origin is a fugitive and not a refugee. Evolving from the legal definitions, the fundamental characteristics of an African refugee are that:

i) the person is outside the country of his nationality, that is, he/she is not a refugee until he/she has crossed an international border;

ii) the person is unwilling or cannot, for the time being, return to his/her homeland because his/her freedom or personal security would be at risk there; and

iii) the persecution he is afraid of is directed against him personally or against a group to which he belongs; or he is indirectly affected by some internal conflict or attack on his country.

The different concepts so far attributed to the refugee suggest that the issue of identity of the refugee is more complex and more controversial than it first appears.

Other Concepts

There are other concepts of 'refugee' propounded by various organizations, such as the All African Conference of Churches, in association with the World Council of Churches; the Inter-governmental Committee for European Migration (ICEM); the Cartegena Declaration of 1984 of the American States; and the mass media.

The mandate and sense of mission of the churches to serve all mankind impels them to extend their concern beyond the category of refugees defined by both the UN Refugee Convention and the OAU Refugee Convention. Consequently, refugee programmes of the churches embrace any displaced person in urgent need, arising from any kind of tragedy; or any victim of society, irrespective of state consideration (AACC, 1971).

The ICEM contends that a 'refugee' is any person who 'has been the victim of war or a disaster which has seriously disadvantaged his condition of living'. Thus the ICEM concept is broader than the legal definition (ICEM, Geneva).

The Cartegena Colloquium of 1984 takes into consideration the refugee situation in the American region where the mass movements of uprooted people were in response to hostilities in the region, and not essentially in response to persecution of individuals. The definition of refugee contains the elements of the UN Refugee Convention (1951) and the protocol (1967) and adds other dimensions of general conflict to include:

> persons who have fled their country because their lives, safety or freedom have been threatened by generalised violence, foreign aggression, internal conflicts, massive violation of human rights or other circumstances which have seriously disturbed public order.

The mass media employ a much broader approach to the definition of refugee by considering any person who is forced out of his home by natural disasters such as earthquakes and flooding; by man-made disasters, including all kinds of conflicts, related or unrelated to government, as a refugee. For instance, the mass movements within Afghanistan, Angola and Mozambique during the periods of conflict were referred to as refugees. Similarly, people in forced relocation programmes, and Ghanaians returning home after being expelled from Nigeria were all referred to as refugees.

Refugee Status

The International Refugee Convention requires that persons needing the protection of governments other than their own, that is, persons seeking asylum, ought to acquire Refugee Status in order to stand apart from other immigrants who do not need such protection. The special status of the refugee poses certain legal obligations on governments that recognize him. In view of these legal obligations, particularly as governments are the sole providers of protection, the International Refugee Convention gives governments the prerogative to determine refugee status.

The OAU Refugee Convention stipulates that Member States should endeavour to grant asylum to those who seek it, Hence, contracting states are obliged to grant asylum-seekers the means of acquiring Refugee Status. The

Universal Declaration of Human Rights (1948) states that: 'Everyone has the right to seek and to enjoy in other countries asylum from persecution'.

This declaration endorses the need for governments to grant asylum-seekers Refugee Status as guarantee for their protection from persecution.

Refugee Status is granted to individuals according to set criteria, the basic elements of which are clearly specified in the definition of refugee as contained in the UN Refugee Convention. Granting Refugee Status is therefore Declaratory, that is, it states the fact that the person is a refugee and should be treated as such. Where there is a mass movement of rural folk, determining eligibility on individual basis is not practicable, hence, a *prima facie* eligibility, that is, 'eligibility based on first impressions' is applied, and all are granted refugee status 'en masse'. However, any individual among the masses who gives cause for suspicion is singled out for screening to determine his status.

The different definitions of refugee adopted by the various sectors and the social, economic and political factors influencing decisions for recognition of persons as refugees have resulted in different categories of refugees.

Convention Refugee Status

This is granted by governments party to the 1951 Refugee Convention, and it confers on the person so recognized a number of economic and social rights as specified by the Refugee Convention, These include the right to obtain Convention travel documents, a protection issue of vital importance. It is also a guarantee against refoulement where the refugee cannot be returned to any place where his life will be in danger. From the refugee's point of view Convention refugee status is the most favourable.

Humanitarian Status

This status is applied by some governments to persons they do not consider qualified for Convention refugee status, but who would be in danger if returned to their country of origin. The refugees have exceptional leave to remain in the country of asylum and also benefit from the principle of non-refoulement. They cannot, however, make claims to benefits in the same way as Convention refugees. A church may undertake to care for such persons.

Mandate Refugees

Mandate refugees are persons recognized as refugees by the UNHCR by virtue of its Statute. In this instance, recognition of persons as mandate refugees does not depend on whether the state of asylum is party or not party to the International Convention. They may even be persons whose applications for recognition as Convention refugees have been rejected. UNHCR operates at different administrative levels in every country where there are refugees. It participates in procedures of identifying persons as refugees, and in this capacity takes responsibility for the asylum-seeker not recognized by government who they think falls within their mandate. A person recognized as a mandate refugee will benefit from international protection against refoulement, and be assured of treatment in conformity with basic humanitarian principles, It does not, however, imply the same social and economic entitlements as accorded to Convention refugees.

Wider Definition

Persons applying for refugee status may not be able to claim a 'well founded fear of persecution', yet would risk danger if returned to their country of origin. Such persons may also be recognized as refugees in a wider sense than the statutory definition. They are treated according to basic humanitarian principles and also to non-refoulement.

Exclusions and Cessation Clauses

The 1951 UN Refugee Convention specifies the circumstances in which refugee status does not apply, or ceases to apply. The exclusion clauses stipulate that the Convention does not apply to:

a) persons receiving protection or assistance from organs or agencies of the UN other than UNHCR like the Palestinian refugees within the area of operations of UNRWA;

b) persons benefiting from the same rights and obligations as nationals of the country in which they have taken residence; or

c) persons who have committed a crime against peace, a war crime or a crime against humanity; a serious common law crime prior to admission to the country of asylum or an act contrary to the purposes and principles of the United Nations.

The cessation clauses stipulate that a person can no longer be considered a refugee if, for example, there has been a fundamental change of political circumstances in the country of origin enabling him to take up renewed residence there.

An essential difference between the exclusion and cessation clauses is that the exclusion clause is applied during the period refugee status is being determined. The cessation clause, on the other hand, is applied after the person has been recognized as a refugee.

African refugees flee from war ravaged situations and violence, from oppressive political conditions or from racist regimes. It is obvious that the refugee is a product of the socioeconomic and political conditions which are characteristic of the contemporary world.

3 Life in Refuge

The moment an individual decides to leave his country of origin to seek refuge in another country, he places himself in a vulnerable situation. The journey into refuge and life in refuge are generally fraught with serious problems, the nature of which needs the attention of governments and the international community.

Uprootment Crisis

People are forced to leave their homes, familiar environment, friends and relations, occupations, established social services and all the comforts of their country of origin, however minimal, to face an unpredictable future which holds all sorts of dangers.

An individual or a whole community might take a rapid, unplanned decision to flee the country, as an anxiety reflex or a crisis syndrome, in response to some major disorder in the country where some tragedy is closing up on them. Sometimes the flight is quickly planned, individuals taking time to pack a small bundle of clothes and food and collect any reserves of cash from home or bank.

The most ingenious methods of escape are devised by people fleeing their country of origin. They often escape during the night, the hours of darkness offering the best prospects of a successful arrival at a place of safety. Those escaping from pursuers evidently avoid all normal routes and direct transport, the route taken being determined by consideration of safety to avoid the enemy. Several routes may be tried in an attempt to find the shortest possible detour, after learning that one way or another has been cut off by the enemy.

Night-hikers penetrate bush or forest on foot, scale mountains or cross rivers and streams on their own, irrespective of weather conditions. More often than not the women carry basic necessities on their heads, with babies or toddlers tied to their back or breast and young children running alongside as best they can.

The duration of the flight varies according to the distances involved, the topography of the area, and the physical condition of individuals, and can mean as little as a few hours' walk to as much as a month's journey.

A fleeing community is at times unsuccessful with its plan of action, the enemy at times catching up with it. There is the possibility of military attack or police arrests before they are able to secure refuge. Some individuals suffer other calamities on account of the terrain. They drown in rivers while sailing in overcrowded canoes or are devoured by wild animals while in hiding in the forests. Having endured a long, difficult and dangerous journey, their immediate needs are the very basic ones – food, water, shelter and medicines. In a situation where a massive exodus is concentrated in a small area, these basic human needs become very scarce predisposing them to starvation. Whether their journey is by rivers or by land, the people arrive at their final destination at international borders physically exhausted and in extreme emotional stress. Most have encountered life-threatening situations and some have had the misfortune of burying their loved ones en route.

The creation of international borders and the presence of security personnel at border posts serve as a deterrent to arriving asylum-seekers. People take the risk crossing the borders at unauthorised points and mingle quickly with the nationals. At the border posts they are often held at bay pending instructions from authorities to security officers to open the gates and let them in. Others are unable to cross the border on account of political constraints between their country and the potential country of asylum, and settle in border areas till they gain admission.

Thousands of people fleeing the civil war in the Horn of Africa in search of refuge travelled a long distance across the desert, and being exhausted, settled in a hilly district. Water was scarce and sanitation was poor; certain death was imminent. Then came a sudden downpour of rain to fill the ditches. It was a blessing. They scooped the water and quenched their thirst, but the water was highly polluted with human waste. A cholera epidemic ensued. By the time international aid reached them many thousands had already died. Management of the dead was extremely difficult, thus worsening the already precarious situation. The survivors were a health risk to the neighbouring communities in the country where they were seeking refuge. Quarantine measures were undertaken immediately, and treatment of the sick and mass immunization started, but the damage had already been done – through uprootment from the country of origin.

An individual attempts a voyage by sea or by air to another country carrying false travel documents or none at all. He hides or poses as someone else, and is in peril till the journey is over. His arrival at the port is met with resistance from immigration authorities and he faces a penalty if he tries to enter the country illegally.

Individuals found on board ocean liners are either thrown into the sea (*Refugees*, No. 27, March 1986, p. 34) or are imprisoned on reaching their destination. Mozambicans who entered South Africa as a result of fighting between government forces and South African-backed rebels in Mozambique, about 15,000 persons, were considered as illegal immigrants. When caught, they were taken to the border post at Komatiport and sent back to Mozambique to face the enemy. Between 1,200 and 1,500 were forcibly returned every month (ibid.).

The fact remains that fleeing one's country from any cause, and seeking refuge in any form, carries so much risk and is an ordeal only a few are able to endure. In our present era of development in Africa, uprootment frustrates efforts at enhancing the living conditions of nationals. It wastes resources as this applies both to refugees who lose all their possessions, and to the host community that has to share limited resources with them. Uprootment adversely transforms life; it disturbs the mental faculties and shortens life expectancy. Uprootment is a crisis situation, indeed a human tragedy; yet people are forced to take such a decision.

Problems in Acquiring Refugee Status

The first and most urgent need of the asylum-seeker is to legitimatize his stay in the country in which he has sought refuge by acquiring Refugee Status. The status is conferred on him by the state that has the ultimate right to determine whether or not he is eligible for the status. Eligibility criteria include verification of the applicant's identity; that he is not a security risk to the country of asylum. Particularly, he must be able to convince the authorities that his life will be in danger if he returns to his country of origin – the persecution factor. Eligibility tests can be influenced by nationality and the absence of identity documents; by social background – religion, culture, occupation – military or political. It can also be influenced by the person's economic status whether economically independent or impoverished and needs help. Others are the mode of entry into the country of asylum, language difference and difficulty in expression. Those who fail the test may be subjected to a great deal of anxiety.

Sometimes a temporary status is granted pending the result of an appeal made to a higher authority. Meanwhile, there is surveillance on the person's movements, associations and contacts. Should he get into any kind of trouble he is quickly detained by Security Officers.

Some countries take a narrow view of their conventional responsibilities to asylum-seekers restricting even the access to an eligibility test. Some limit the enjoyment of the applicant to the normal right of residency during the period of screening by detention and jail terms. In some instances, the asylum-seeker's military status alone results in his outright refoulement. Some countries fine the airlines and shipping lines for bringing in persons carrying inadequate travel documents and the asylum-seekers are returned to places of embarkation at the expense of the airline.

Lack of access to appropriate eligibility-determining procedures is a pressing concern. Administrative procedures at border posts have, in some countries, replaced proper procedures for screening or determining status and providing adequate legal guarantees. According to records, there are instances in which asylum-seekers denied admission into the country without the benefit of due process of law were subjected to particularly severe punishment on return to their country of origin.[1]

Moreover, the expansion of visa coverage by some countries specifically to embrace refugee-producing countries, coupled with the strengthened mechanisms to enforce immigration controls, effectively close entry to a number of asylum-seekers. While such immigration controls are directed at immigrants generally or abusers of the asylum process, they work indiscriminately to hinder access of genuine asylum-seekers to eligibility procedures and to the rights and protection they are entitled to. However, people having economic or social difficulties do sneak into countries of asylum. When their identity comes into question they face problems in gaining a status.

As much as refugee status guarantees the refugee the rights he so urgently needs to adjust to a new life, there are circumstances where an asylum-seeker would rather not accept refugee status. Some feel embarrassed at being defined as a special kind of individual – a feeling of being 'branded'. Some experience fear that by establishing a fact of persecution against the government of the country of origin they become traitors. On the contrary, some refugees use their status as an excuse for clinging to other refugees and avoiding contact with host nationals. Still others use their status as a means of drawing attention to themselves and their life situations.

A group of nomads searching for water in severe drought, or persons displaced by natural disasters, may secretly cross a national border in a remote rural area into another country. Once the need is satisfied and prospects for the future are bright, such a group may refuse to return to the country of origin. Ethiopian refugees entering northern Somalia and Djibouti are an example. They were fleeing the war in the Ogaden desert between Somalia and Ethiopia, but more importantly, they were fleeing drought.

Aliens legally resident in a foreign country whose resident documents expire, but who on account of disorders in their country of origin which may affect them personally, are unwilling to return home or to renew their identity documents, have to seek refugee status. If their application for Refugee Status is rejected, they face residence problems.

Protection Problems

When an ordinary alien abroad comes into conflict with the law, he can turn to his country's foreign mission for help. The mission intervenes on his behalf by approaching a government agency if it appears that its national is being treated in a discriminatory manner or that his case is not being dealt with under due process of law.

A refugee who finds himself in trouble with the authorities of his country of asylum has no such recourse. He has declared that he has no confidence in the protection of the government of his country of origin because his government has a resentment against him or has wicked intentions against him. In essence, it was upon these same grounds of lack of trust that he gained refugee status.

A refugee or an asylum-seeker accused of a crime may be detained pending an investigation of the circumstances. A situation such as this is termed 'justifiable detention'. Unfortunately, however, there are instances where an asylum-seeker has been unjustifiably or arbitrarily detained because he entered the country in an irregular manner.

A person's detention under any circumstances has particularly grave consequences. The fact that he is a refugee or an asylum-seeker and therefore without the protection of his country of origin makes him liable to discriminatory treatment. He also runs the risk of being served with an expulsion order and even of being forcibly returned to his country of origin.

The cases of Ms A and Ms Y illustrate instances of denial of protection for refugees and asylum-seekers. They are drawn from UNHCR records,[2] but for the purpose of confidentiality neither the persons nor the countries concerned are disclosed.

Case of Ms A

A refugee woman was arrested on suspicion of being a subversive. She was detained without charge under a state of emergency. Her release was made subject to her expulsion from the country to a second country of asylum. On three occasions the expulsion order was denied her even though Ms A had been granted the necessary visa to settle elsewhere.

Ms A contracted tuberculosis and her situation preoccupied human rights circles in a number of countries. UNHCR made continued demands to secure her release. Finally, success was achieved and after six and a half years in detention, still without charge, Ms A was released from prison and permitted to resettle abroad. She left behind a child whose whereabouts, despite efforts to trace her, is still unknown.

Case of Ms X

This young woman and her child were detained by border security officers of the country of asylum then thrown into prison. She was part of a group of asylum-seekers who were suspected of being spies or subversives and who were routinely detained at the border, Some were released after a few months once their bona fide identity had been established. Others, like Ms X, remained in detention for an indefinite period. UNHCR learned of her predicament quite by chance. By this time, Ms X had been in prison for some five years without charge or trial. Finally, the authorities agreed that both she and her child, together with her other compatriots, should be transferred to a refugee settlement. This was, however, only a temporary measure and UNHCR was obliged to find resettlement opportunities for the group out of the country.

Friction between refugees and nationals of the country of asylum has resulted in various incidents, much to the disadvantage of refugees. Some nationals resent the allocation of fertile land or fishing rights to refugees, while others resent the idea that national resources are channelled to the care of refugees. At the height of their discontentment, they flair up against refugees. Nevertheless, once a person has been recognized as a refugee, it is assumed that he can settle in the country peacefully.

Some refugees have had genuine fear for their personal safety in countries of asylum. They have been attacked by assassins who sneaked into their places of residence or mingled with large refugee groups; and they have been bombarded by military agents from their country of origin. Some refugees have been conscripted at gunpoint to serve in rebel forces fighting government forces in countries of asylum. They have had to carry out a variety of tasks including carrying equipment and food, planting mines and participating directly in fighting.

Forced conscription of refugees, for short or long periods, into the rank and file of warring factions is a particularly complex problem. Not only does this practice seriously endanger the lives of those directly involved and their families, but it also jeopardizes their refugee status and compromises the humanitarian and civilian character of the settlements in which they live. Although refugees are under legal obligation not to get involved with armed conflict,[3] they were used under duress during the Congo crisis. Two UN officials protecting them with the UN flag were killed. Such an incident has happened elsewhere in Africa where refusal to fight had resulted in the refugees being executed by the warring factions. Unfortunately, this is a continuing practice which requires considerable effort to halt.

In other instances, conscription into warring factions has been voluntary when refugees are facing starvation and have been promised food or money. In some other instances, such a promise is to facilitate the acquisition of citizenship once victory has been achieved.

It is certainly no fun being a refugee in a settlement, whether or not it is enclosed with barbed wires. A refugee settlement is sometimes attacked by assailants using rifles, sub-machine guns and knives, or it is bombed by aircraft from the country of origin, the settlement becoming a death trap.

Persons pursued by an enemy develop a fear of the persecutor with a resulting tendency to develop anxiety disorders when the danger is over. There is intensification of the instinct of self-preservation and deterioration of moral values. These disorders affect others in the settlement. Some refugees, who have been attacked several times, are unable to cope with the military atmosphere where government security forces, and even some of their menfolk, participate in surveillance duty over the refugee settlement carrying arms.

During the period the liberation movements of Zimbabwe were struggling for independence in the late 1970s they took refuge in settlements in Mozambique and Zambia. Rhodesian mercenaries attacked them by air, landing by parachute, and killed whole communities. Schoolchildren and their teachers, the catering staff, medical staff, inpatients and even others being conveyed by ambulance to hospital perished. The victims were not guerrilla fighters; they were civilians. They were buried in mass graves. The assailants were also suspected of attacking the administrative staff of the refugee communities in the towns by sending them letter-bombs.[4]

The following is a sordid account of a military incursion over a refugee camp in a country in southern Africa.

> On the day of the attack, a refugee woman was so panic-stricken and so mentally confused that she ran into a pit latrine and unknowingly fell into the pit. She was immersed in the contents of the pit up to the level of her chin. Her survival lay in her raised arms which held to the sides of the pit.

The incursion was swift and deadly; all the other residents were murdered; she was the only survivor in that section of the settlement. She was rescued by the national army which chased the attackers out of the territory and rushed to the settlement looking for survivors. When she was brought out of the pit she had lost her senses and remained irrational for months in hospital.[5]

Human beings who have already risked their lives seeking refuge are once more subjected to such a calamity in flagrant violation of all the moral standards of our civilization.

Below are other instances of mass attack on refugees (*Refugees*, No. 4 August 1983, pp. 5 and 11):

a) 4 May 1987, in Kassinga, Angola: 600 Namibians were killed and 400 injured by air bombardment;

b) 30 January 1981, at Matola in Mozambique: 12 refugees were killed, and seven seriously injured by machine-gun fire;

c) 9 December 1982, at Maseru, capital of Lesotho: 42 persons, of whom 30 were South Africans and 12 were local inhabitants, were killed by heavy artillery fire;

d) 23 May 1983, Mozambique, 10 kms from Maputo, Mozambique: six people were killed, including Mozambican and South African civilians; and 40 injured as a result of aerial bombing.

One of those who escaped the bombing of the Kassinga settlement by the South African planes described what happened (ibid., p. 18):

> That day, I got up at 6 o'clock in the morning as usual. After breakfast my comrades and I went across singing towards the large tents in the middle of the

settlement which we used as a school. It was 7 o'clock and the bell had just
rung for the beginning of classes. We were getting into line when all of a sudden
there was noise in the sky. Airplanes were dropping things. We thought they
were bringing us food. What a tragic mistake! The objects exploded when they
hit the ground. I escaped into the nearby bush, My friend had a leg broken.
Someone pointed out to me that I was injured. I was bleeding profusely in the
right hand but I was not at all afraid. In any case I felt I was already dead.

As a result of this attack there were 600 killed, 400 wounded, 100 or more
buildings and huts destroyed, including hospitals, schools and storage depots.
The victim quoted above lost all the fingers of her right hand and her arm was
severely mutilated. Sometimes she showed signs of emotional disturbance
with crying episodes, timidity and withdrawal into herself.

Apart from these brutal attacks there were numerous kidnappings and air-
raids throughout southern Africa. Similar incidents occurred in the Horn of
Africa where several refugees were abducted by agents from the country of
origin.

Problems of Assistance

The management of an exodus demands sincerity, devotion to duty, and
integrity on the part of the support system. It also poses a challenge to the
patience, understanding and cooperation of asylum-seekers.

Emergency aid problems concern particularly sizeable groups of asylum-
seekers usually originating from rural areas who have not been able, in view
of their precipitated flight, to take with them enough food and clothing. To
these asylum-seekers the problem of survival is paramount, and it is most
important to provide them immediately with emergency food, water, temporary
shelter and medicines.

They have to accept any kind of food supplied; not always the kind to
which they are accustomed. Food is rationed and meal times regulated.
Refugees sleep, often overcrowded, under temporary shelters on mats placed
on the floor. They have to get quickly accustomed to being sent around to
carry heavy sacks or bundles of food and other supplies, to distribute food
and clothing, carry sick people and children, or dig graves and bury the dead
– chores some of them have never carried out before. They must remain on
the alert to identify potential enemies should they have any fear of being
followed.

They have to cope with any sudden relocation movements of their group for settlement, and travel with familiar or unfamiliar people. Their minds may not be at ease because they cannot find close relations who may be lost, dead or left behind in the country of origin. Emergency movement of a large group poses problems of transportation and difficulty in organizing the group by families, village groups or ethnic groups. This operation when properly executed solves adjustment problems when they finally settle. Refugees cling to their traditional way of life whenever possible and feel safe when moving together with kith and kin rather than with unfamiliar people.

Sometimes the emergency aid provided is in conflict with the economic and social situation of the local national population. This leads to discontentment of nationals which impairs the future harmonious settlement of the refugees in the country of asylum. Other times it is so liberal that it discourages the initiative of the refugees and makes them think it is the normal way of life in refuge where nationals are always at their beck and call.

Problems arise when it is considered unsafe, for health or political reasons, to settle refugees at the border areas. When refugees have strong feelings of retaliation against the government that turned them out of their homes, they take advantage of living close to border areas to carry out acts of aggression against the country of origin. This attitude becomes a source of conflict between the country of asylum and the country of origin. The OAU frowns over this development and requires that refugees should not be settled around border areas, and their presence should not be a source of conflict between Member States. Consequently, it has become standard practice in countries of asylum to relocate refugees long distances away from the borders as soon as emergency aid procedures are over.

Problems with Adjustment

Warm reception and hospitality by the host community are crucial issues. The size of the influx in hundreds of thousands within a short space of time – within a week – exasperates the national community and creates a variety of hostile attitudes against them. Differences in ethnic, cultural and religious backgrounds may also influence reception. Even where asylum-seekers and their hosts are of the same ethnic origins they do not fare any better. The initial show of ethnic solidarity wanes as soon as food shortages occur. Discontentment becomes obvious among nationals and once again the new arrivals are evicted from their lodgings to seek international assistance. Thus humiliated, refugees develop a sense of bitterness and regret ever leaving their country of origin.

It is a popular theory that fear or distrust of things that are foreign is a fundamental part of the human psyche. In difficult times, this latent instinct is unleashed and it takes on the dimensions of a phobia. Recent manifestations of xenophobia (*Refugees*, No. 4, May 1984, p. 20) – a mental state of resentment of foreigners – is widespread in societies of countries of asylum. A fear of foreigners becomes a fear of refugees. The fact that the refugee's image is tarnished by the causes of his uprootment and his dependence on countries of asylum for sustenance, this factor weakens his efforts at adjustment.

While the general attitude of Africans is to receive aliens with courtesy and warmth, there are always significant exceptions where refugees are concerned. They are resented and shunned, stigmatised and humiliated subjected to indignities and discriminated against.

Urban Refugees

Individual refugees, particularly youths, seeking educational opportunities, turn misfortune to blessing and make every endeavour to adjust to the new environment against all the odds and pursue their studies. However, some youths have the tendency of over-evaluation of the country of asylum or person in authority. If their expectations are met, in course of time they adjust. But should they experience hardship – the deprivations, discriminations and rejections – they become frustrated. This feeling advances to aggressive behaviour. Such negative behaviour has surfaced on a number of occasions when individual urban refugees needed assistance. An unfortunate incident occurred in Ethiopia where refugees, angered by the inadequacy of monthly allowances offered them, attacked their supervisors, killed a lady and wounded several others.

There was a young urban refugee who acquired a positive attitude to his rather stressful experience when seeking refuge. He said he was a school teacher and a national of country A in southern Africa. He went on holidays to visit his uncle in country B across the border. A couple of weeks before the holidays were over border clashes between countries A and B erupted. He tried to return home by bus before things got worse. Soldiers stopped the vehicle and abducted all the young men. He was conscripted into the army of country B. His plea that he was not a national of country B was not accepted because he spoke their language fluently. Months later, while on military operations along the border, he escaped into the bush. He hid his gun, stripped off his uniform, and half-naked, managed to walk as far as the nearest village into country A. Soldiers from country A spotted him, arrested him and called him a spy. He was subsequently detained in his own country. Fellow ethnics in that military unit, hearing him speak their local dialect, helped him to escape. Together with a large exodus of countrymen heading towards country C to seek asylum, he entered country C.

He joined an ethnic group and was granted asylum without incident. He vowed never to return home. He hoped his misfortune would turn to blessing, and in spite of the severe panic he suffered in his encounters, he remained emotionally stable.[6] He is one of many who needs understanding and support to prevent an emotional illness.

Some urban refugees, frustrated over their inability to make satisfactory adjustment to their new environment, develop minor psychiatric disorders. Some become drug addicts. Others who have not been able to overcome the crisis of the initial encounter with disaster in the country of origin and also have anxieties in adjustment, develop psychogenic disorders, including ideations of persecution, antisocial behaviour, as well as varying forms of psychosomatic illnesses. Medical care for such refugees is very costly.

It is expected that once refugees have gained a status they should make every endeavour to take advantage of any material assistance available to adjust to their new environment. More importantly, problems of adjustment pertaining to differences in culture and traditions between refugees and nationals should normally be overcome through integration. Adjustment takes time and is a challenge, particularly to those who have had painful experiences.

Vulnerable Groups

In every society are vulnerable groups of all ages, male and female – who need special care and attention because they have peculiar problems which expose them to exploitation by other members of the national/refugee society. Refugees form one such vulnerable group because of the circumstances of uprootment and nature of their needs. Those refugees who are in a specific precarious circumstance have double calamity. These are the children, the orphans and the unaccompanied minors; the elderly and the disabled; separated families and others in distress.

The Family

At the initial stage of uprootment during a disorder in the country of origin, people are displaced. Separation may be deliberate when, for instance, a group of refugee children are taken for protection in neighbouring countries, as it happened during the Nigerian civil war. These children were later returned and found a home. Separation which occurs arbitrarily during uprootment is of special significance. Parents become separated from children, spouses find themselves on different sides of a border. They may or may never get reunited.

In all countries and cultures people have a basic instinct of belonging to their immediate household and to their extended family. In the case of refugees, particularly Africans, this instinct is very strong. Loss of family contact is traumatic, especially for children. In most cases, the scars of separation can never be healed, even if the refugee families concerned have been able to reunite and return home or begin a successful new life in refuge. This state of affairs can be ameliorated if ethnic groups in refuge are allowed to live together, at least, for moral support.

In a refugee family roles change. Women may find themselves as heads of households, undertaking activities and assuming responsibilities which are new to them. Men have to come to terms with a change in their traditional status and participate in surveillance duties. Assigning to refugee children numerous responsibilities unrelated to their age constitutes child abuse. They may, however, adapt more quickly to the new way of life, but this causes a lot of pain for parents who wish to safeguard their culture. In many refugee settlements, frustration and stress lead all too often to violence within the family.

Children

Children suspected as orphans are taken to an organized settlement for care. There is no opportunity for adoption as their parents may or may not be alive. Should adoption be possible, it is not recommended till repatriation is effected. Where there are very large numbers of refugee orphans in a settlement they are camped together in the care of a 'mother'. These are the children who suffer the most psychologically because should repatriation be delayed, they grow up to adulthood without a proper home upbringing and association with adults. They have no childhood experiences of family life or even of the freedom of interaction with the world at large.

It is difficult to express all the losses that a refugee child can suffer. Aside from being orphaned and lacking extended family support, children are spectators of disease, torture and death of which they become a part sooner or later. They are victimized and exploited by adults, yet their voices are not heard; only their cries can be heard with little sympathy.

Hundreds of refugee children with different degrees of physical disabilities, acquired in conflict, were assembled in a city for a musical performance. The disabilities were so overwhelming that some spectators, filled with grief, were emotionally upset.

Children accompanying adult groups often suffer a variety of deprivations such as food shortages and irregular meals, lack of education and poor health facilities. Many die from outbreaks of diarrhoea, measles, and meningitis. Many are maimed by poliomyelitis because vaccination is delayed, or because their resistance to infection is lowered by malnutrition.

Even the fortunate refugee children whose immediate family remain intact do suffer. Fathers are absorbed into vigilante groups to protect the refugee territory. Mothers work in communal kitchens. The children fetch water from distant streams or collect firewood in snake-infested forests. Generally they work to meet survival needs and seek moral support from one another.

Children brought into refuge soon grow up and without the necessary parental guidance or supervision by older members of the refugee population adopt negative attitudes to settlement life. Without adequate education and skill training they join the ranks of the unemployed and become a social burden on society. They may not be aware of their status as refugees, or fail to appreciate that as refugees they have legal obligations to respect the laws of the land that has given them refuge. Once caught indulging in any antisocial activities, law-enforcement agencies take severe punitive measures against them, up to imprisonment. The criminals are pushed out of the country. Families are unable to trace them until UNHCR becomes aware of an arrest and intervenes on their behalf. In spite of this, some refugees disappear under mysterious circumstances. A large population of refugee youth is a security risk to the country of asylum by virtue of their idleness and desperation to survive.

Meanwhile, without effective rehabilitation measures, the integration of youth into the wider ethnic community becomes prohibitive. Moreover, if their stay in refuge is prolonged, their assimilation within the wider national community proves difficult. This is particularly important should they contract marriages with nationals.

Women and Girls

Females constitute the majority of Africa's seven million refugees. A rural exodus normally consists of a greater percentage of females, the majority of whom are married and have children. Most are from farming or fishing communities. They face the same hazards of the journey into refuge, similar problems of acceptance by host nationals, adjustment problems and a variety of deprivations and discriminations.

The peculiar problems of females throughout the ages, and in almost all societies of the world, are sexual abuse and physical hardship. Rape has occurred as a torture method to acquire information, as an act of aggression to subdue victims of a conflict, and as a condition in exchange for favourable treatment at national frontiers and in refugee settlements. In some countries rape is punishable by long jail sentences or by the death penalty, but when a refugee female is the victim, punitive measures are less aggressively pursued.

Paradoxically, female refugees, much like other females in extreme economic need, indulge in indecent behaviour, such as prostitution. In a refugee settlement, young attractive girls sometimes expose themselves to this practice to earn extra money and make ends meet. When some of these girls contract a sexually-transmitted disease, the infection usually spreads among the refugees, and medical care is difficult to come by.

It is not surprising, therefore, that a high birth rate is reported among refugees even when they are under the threat of starvation. Teenage pregnancies are common partly because of the vulnerability of the girls, and partly because of the lack of opportunities to enhance their personal well-being.

While some male refugees have to submit to degrading work in order to obtain the means for supporting themselves and their families, the females are sometimes grouped alongside the males for work, irrespective of the physical energy required to accomplish it. Forced labour for females is undignified and stressful.

A female family head normally experiences economic hardships in maintaining her family. As a refugee, she suffers the emotional stress and physical hardships as well. She is allowed very little parental control over older members of her family. Sex discrimination is often apparent as in the unequal participation of women in the design and implementation of refugee assistance programmes.

Elderly Refugees

By virtue of age and diminished physical abilities, elderly refugees are exempted from various assignments allocated to members of a large group. When the day begins and workers are out of sleeping quarters, the elderly resign themselves to an isolated life. Nevertheless, they occupy a place of influence over the younger members of the group, and some remain as heads of ethnic groups. Their period of isolation is often used to reminisce and to consider strategies for controlling the group socially. In some cases, however, such influence or social control is prohibited and the elderly are kept segregated.

Sometimes they are considered insubordinate to settlement authorities for exercising traditional or social control over the group, and are also held responsible for acts of indiscipline among younger members of the group. Psychological problems stem from their inability to put their experience at the disposal of authorities on matters affecting the group and to their forced isolation.

Elderly refugees are unable to learn a new language easily, adapt to unfamiliar people, or accept rigid orders conducting their lives. An inner conflict resulting from humiliation causes rapid deterioration of body and mind.

While facing the common hardships of subsistence such as poor housing, insufficient clothing, inadequate health care, discriminations and humiliation, the elderly also face the usual health problems associated with ageing such as organic brain syndrome and arthritis. They need attention and emotional support and are generally unable to adjust to the changed environment. This, coupled with nostalgia of country of origin, and the thought of dying and being buried in a foreign land, can be very stressful indeed.

Disabled Refugees

The disabled have been with mankind throughout the ages and will continue to live with society in times of peace or conflict. Disability in the developing world has commonly resulted from poliomyelitis, organic brain damage, and from injuries. Uprooted populations have seen war and violent conflict in their country of origin and also in their country of asylum and many have become handicapped. Some have lost their sense of hearing through deafening noises of firearms or lost their sight in severe burns. Others have been severely mutilated by fragments of exploding bombs and lost their arms or legs or both. The fact that refugee communities constitute a substantial number of disabled persons is not at all surprising.

Under normal circumstances, disabled people represent about 7 per cent of the population in a developing country (*Refugees*, No. 66, July 1989, p. 20). In a situation of conflict, however, that figure rises to about 10 per cent (ibid.). The Angolan conflict led to the mutilation of about 10,000 people in 1985 alone (*Refugees*, No. 4, August 1983, pp. 11–18). In Mozambique thousands were disabled not only in attacks by Renamo armed bandits, but also as a result of malnutrition, untreated diseases, lack of immunization and inadequate health facilities.

The modern trend is to educate disabled people and absorb them into the economic mainstream. Enabling disabled refugees to play a full and fulfilling role in their community is not an easy task. It is expensive, time-consuming, and easily overshadowed by more visible social needs. In the absence of constructive activity and opportunity for social integration, the disabled and the aged alike are neglected – a waste of human potential.

Rebuilding one's life in a strange and unfamiliar environment is a long and painful struggle. After the shock of the initial trauma has faded, the horror and disorientation associated with the loss of family, community, culture and hope continue to weigh on the refugee. The capacity of the refugee to adjust to his new way of life is even more traumatic. It is the human support system of friends, family and community which provides the ingredients for emotional healing.

Problems with Education

Refugee education in the country of asylum poses a problem in view of the large number of children in a refugee population. Nevertheless, elementary education is usually obtainable. The schools need not be permanent structures; the main inputs are facilitators with some educational background and adequate basic teaching materials. Beyond the level of primary education more funds are required to meet the cost of scientific materials, more durable structures for assembling equipment and remuneration for well-trained teachers. Many countries are unable to provide these adequately, even for nationals. In the Sudan, for example, young refugees, many of high school age, desperate to further their education cannot do so. They are restricted by inadequate facilities and the inability of local educational institutions to absorb them. At the tertiary level, refugees have to compete with nationals to obtain scholarships.

Problems with Employment

Third World countries, and even industrialized nations, face serious unemployment problems. Africa is in the process of industrial development, agricultural diversification and expansion, hence has a greater problem. Rural refugees manage on their own, circumstances being favourable, to practise their professions in agriculture, handicrafts, village industry, etc. Urban refugees such as businessmen, doctors, teachers and engineers must obtain work permits to seek employment. But in the main, they have to compete with nationals to secure a job. Registration for an employment permit is as demeaning and as bureaucratic and lengthy a process as obtaining refugee status.

Refugees arrive in urban areas of countries of asylum with occupational skills and experience, sociocultural values, ideals, and beliefs and norms, as potentials which sometimes do not correspond to or fit into the prevailing culture or mode of production. Their integration into the economic sphere depends on their efforts to adjust. Some refugees have expectations that are unrealistically high. After a long period of waiting to obtain refugee status they find job opportunities a nightmare. In Somalia, for example, professionals of this sort languished in towns living on meagre refugee allowances, frustrated by their inability to find a job. A few refugees in the teaching and medical field meet local demands and are fortunate to be absorbed into public institutions or be employed in a refugee settlement. But the majority of unskilled and semiskilled refugees remain unemployed. Health problems of an emotional nature arise and these in turn create a burden of rehabilitation on those agencies assisting them.

Members of the learned professions and civilian administrators have been forced by circumstances to accept domestic work in private homes and government institutions. Kenya, for instance, does not permit large-scale settlement programmes or allow refugees to own land (*Refugees*, No. 23, November 1985, pp. 14–16). Inevitably, professionals among the refugees have to accept any kind of job or stay idle. Consequently, some refugees are unable to marry, and if they marry nationals, they expect the spouse to support them, as well as the children of such a union. Unable to find a job and literally destitute, once their allotments have expired, church charity or begging in the streets becomes their only resort. But even drawing attention of sympathetic passers-by is not that easy. Professional beggars operating in Nairobi, for instance, have highly elaborate techniques to frustrate the efforts of refugees.

This makes it more difficult for legitimate refugees to plead their cause. If refugees live with friends or fellow countrymen in normal residence they are assumed to be independent and there is the tendency to withdraw their supplementary assistance. Without adequate provision, refugees are unable to integrate easily into society.

Problems with Health

Many countries of asylum are beginning to address the critical needs of refugees for emotional support. About half the complaints presented by refugees in outpatient medical clinics appear to be stress-related. The principal causes of stress include profound differences in culture and social norms between refugees and nationals or refugees of different nationalities living together in a settlement; language and communication difficulties; so-called survivor guilt, for being alive when the majority have died; inability to return home and inability to progress educationally or to find economic security. Some of the health problems are the result of inadequate food and shelter, squalor, unsafe sources of water supply, and insanitary conditions in emergency reception camps. They have also been associated with ignorance and illiteracy, and with crop failure and famine.

In Africa, suburbs of cities and rural areas are yet to get their fair share of the social services, including hospitals. In the settlements, though remote, organized health care is often made possible by health professionals among the refugees and by voluntary agencies. Nevertheless, shortage of medical supplies frustrates efforts at achieving necessary goals. Since hospitals are located mainly in cities long distances away, emergency medical assistance is unattainable in difficult childbirth, severe injury such as extensive burns or in severe infections. Under the restricted conditions some refugees live in, it is not possible to seek specialized health care.

Those suffering from diabetes cannot be stabilized and those suffering from debilitating illnesses such as arthritis have to live with it. The onus lies on the country of asylum to provide health facilities for refugees, but here again, facilities are usually inadequate even for nationals.

Problems with Travel

Documents relating to Refugee Status, refugee passports, identity, and endorsements of application for jobs, loans for private enterprises, scholarships, and so on, do not pose special difficulties as these are prepared free of charge. However, difficulties arise when some refugees have to travel for psychiatric treatment or other serious health problems abroad. They need escorts and this can be very costly.

If a refugee has to stay temporarily in a second country of asylum, the refugee passport must contain a 'Return Clause' – a Convention provision which should enable the refugee to return to the first country of asylum. For instance, a refugee travelling to a second country of asylum for medical treatment should be able to return to the first country of asylum after obtaining treatment. Where the first country of asylum considers that that particular refugee is an 'Undesirable', on account of some misdemeanour he has committed, such a Return Clause is denied him. Consequently, the refugee is unable to travel to obtain treatment. If he attempts to travel without the Return Clause this is considered an irregular movement. If he manages clandestinely to reach the second country of asylum without a Return Clause he cannot return to the first country of asylum after medical treatment, and this creates an asylum problem for him. He has to apply for Refugee Status once again, this time in a second country of asylum. If he has left his family behind in the first country of asylum, his problems become more complicated.

Problems of Refugees in Africa Vis-à-vis those in Europe and Elsewhere

African countries are confronted with serious economic, social and political problems, yet they have in the past five decades given generous asylum to millions of people uprooted on the continent, and continue to do so. In some parts of the world, however, prosperous countries are introducing measures designed to restrict or deter the arrival of asylum-seekers on their territory. There are many reasons for this.

In the period following World War II, Western European states received substantial numbers of asylum-seekers and refugees of essentially European origin who were subsequently integrated into their societies. They were also accepted for resettlement overseas by the traditional major immigration countries, such as Australia, New Zealand, and Canada. This situation changed in the 1980s. To a large extent asylum-seekers reaching Europe were made up of persons primarily from Asia, the Middle East and Africa. A growing sector of European public opinion viewed this movement as unjustified, especially because the persons concerned competed for scarce employment opportunities. The cost of assisting them in Europe was much higher, their prospects for integration there were much lower and the cultural, religious and social values in Europe were alien to them. This perception of the newcomers gave thought for new strategies.

European states are of the opinion that asylum-seekers originating from non-European countries should be protected and assisted in the regions where they originate and should not be allowed to travel to European countries. As a result, European states have adopted restrictive measures in an attempt to stem the flow of refugees.

Owing to the high level of association between European states, action by one state influences that of another, so that as each European state adopts the measure, a wall is raised around Europe to prevent the spontaneous arrival of asylum-seekers from outside the region.

Western countries have kept a watchful eye on their borders, asylum has been granted less generously than in the past, and old currents of xenophobia reappear in a region long considered a cradle of human rights. States which more or less grudgingly tolerate the presence of asylum-seekers on their territory often compensate by making refugee status harder to obtain. There is some evidence of this even in the Nordic countries, currently among the most generous in the world. According to the European Consultation on Refugees and Exiles Group, 'Europe is doing too little, doing it in a disorganized way, and the end result is greater suffering and hardship for people with a genuine claim to asylum' (*Refugees*, No. 47, November 1987. p. 41).

The British government has introduced legislation designed to deter air and shipping lines from carrying asylum-seekers without valid passports and visas into the United Kingdom (*Refugees*, July 1987, p. 51). The Minister for Overseas Development, explaining Britain's international refugee policy, said:

> The best and most hopeful assistance we can offer is help for the efforts of their own governments to bring about long-term sustainable growth and development. It is much kinder to offer a solution for refugees who have left their countries because they are, quite simply, afraid to stay – people who are afraid of discrimination or ill-treatment because of their race, religion or political opinion (*Refugees*, No. 47, November 1987, p. 35).

Since 2 July 1982, the government of Hong Kong, as a dissuasive measure, has been locking up Vietnamese 'boat people' in so-called 'closed camps' as soon as they arrive. This amounts to virtual imprisonment. The 'inmates', who are unable to leave, are subjected to strict discipline (*Refugees*, No. 3, May 1983, p. 21).

There is no doubt that compared with some of the Far Eastern countries which threatened to turn the boat people to an unknown fate in the deep seas, and the countries in Europe and elsewhere that scrutinize every individual case, African governments have much more liberal policies towards refugees. In spite of this open-door policy, the plight of refugees is not easily resolved. The influx of a large number of refugees in the world's poorest nations brings with it a variety of difficulties and places great strain on these nations' inadequate national resources. Thus it erodes the hospitality normally demonstrated at the time of arrival. Life in asylum countries is a predicament to the individual African refugee who, as described earlier, often suffers from want of subsistence, lack of education and few job opportunities. Consequently, unless there is active intervention on the part of the international refugee-assisting organizations, the plight of refugees will remain bleak.

Much has been documented about refugee camps in Europe – including hostile treatment given to detainees and human rights violations accorded to refugees. In Latin America, Asia and the Middle East assassinations, abductions, aerial bombardments of refugee camps by agents from the country of origin are a common feature. This is also the case in southern African countries such as Angola, Mozambique and Zambia.

The problems of refugees in Africa often have a happy ending in the mass voluntary repatriation exercises. Elsewhere in the world, such as in Latin America, refugees have also concluded their periods of refuge with a happy return homeward.

In Europe, however, refugees do not often take advantage of the opportunity of voluntary repatriation. By and large these refugees would have lost their roots and have little traditional affinities to their homeland. Once in the country of asylum they normally want to stay. This has resulted in restrictive measures imposed on refugees entering Europe.

Loss of Refugee Status[7]

Normally a refugee who returns to his country of origin voluntarily ceases to be a refugee. He may, on information received from his country of origin, arrange his own departure or make a formal request to an international organization for material assistance.

Loss of refugee status also comes about principally through application of one of the so-called 'Cessation Clauses'.

These clauses, which are contained, in the UN Convention (1951) and paragraph 6A (ii) (a–f) of the Statute of the Office of the High Commissioner for Refugees, spell out the conditions under which one ceases to be a refugee. They are based on the premise that 'International protection may no longer be justified or required because the reasons for a person becoming a refugee have ceased to exist as a result of changes in the country of origin or habitual residence'.

A strict approach is applied to the implementation of the 'Cessation Clause', motivated by the need to provide refugees with the assurance that their status will not be subject to constant review in the light of temporary changes in the situation prevailing in their country of origin.

Paragraph 117 of the UNHCR *Handbook of Procedures and Criteria for Determining Refugee Status* states:

> 117, Article IC does not deal with the cancellation of refugee status.
>
> Circumstances may, however, come to light that indicate that a person should never have been recognized as a refugee in the first place, for example, if it subsequently appears that refugee status was obtained by a misrepresentation of material facts, or that the person concerned possesses another nationality, or that one of the exclusion clauses would have applied to him had all the relevant facts been known, In such cases, the decision by which he was determined to be a refugee will normally be cancelled.

Loss of refugee status through application of the Cessation Clauses must be clearly distinguished from loss of status as a result of annulment or cancellation. The UN Convention (1951) does not specifically address cancellation. Cancellation of status under the circumstances envisaged in paragraph 117 of the UNHCR *Handbook* follows general principles of law, including that of *res judicata*. According to the principle of *res judicata*, once a matter is judicially determined, that matter may not subsequently be reopened by the same parties. However, in some rare circumstances, a decision may lose its final character when new facts appear, indicating that the decision should never have been taken in the first place.

The circumstances that may call for an exception to the principle of *res judicata* include the following:

a) newly discovered evidence;

b) fraud, including concealment of material facts that there was a duty to disclose; and

c) other misconduct in the proceedings.

Exceptions to the principle of *res judicata* must be approached and applied restrictively. The newly discovered evidence of fraud or other misconduct must, for example, be sufficient material to have affected the outcome. In principle, it also has to be proved that the evidence could not have been discovered earlier, that is, at the time the decision to grant refugee status was made. Moreover, it is usually required that the fraud or misconduct be judicially determined.

It also follows that misrepresentation of material facts must normally be intentional and manifest. If the misrepresentation is not manifest there should, at least, be serious doubts as to the plausibility and credibility of statements made. Furthermore, the subsequently emerging circumstances (possession of another nationality or activity to bring the person under one of the exclusion clauses) must not have been clearly evident or readily discoverable at the time of the first decision. As cancellation of refugee status could have the effect of depriving individuals of their acquired rights it is all the more a decision to be approached restrictively.

Some refugees who have no intention of returning home because of privileges they have in the country of asylum have been responsible for losing their status, and consequently their rights. The fact is that they can no longer refuse to accept the protection of their country because the circumstances which led to their being recognized for refugee status have ceased to exist.

The following cases illustrate the circumstances of cancellation of refugee status.

Three urban refugees were arrested in connection with their involvement in various clandestine activities. The first had been stealing repeatedly from neighbours, the second was seen on several occasions loitering around the military barracks, and the third was seen by villagers crossing the border from their country of origin carrying a heavy load of farm produce to sell.

During interrogation at a police post it was discovered that the first person was not a national of the country of origin. The information he gave concerning his identity in his application for refugee status was not correct. The second had contacts with some nationals of the country of asylum and had been sending information to the government of the country of origin. Indeed, he was not suffering any persecution – the basis of obtaining refugee status. The third had been going in and out of his country of origin for several years and farming in his own village. These three persons were brought before a court and when their activities were substantiated their refugee status was summarily withdrawn.

Conclusion

The crisis of sudden uprootment from one's comforts at home and the inability to legally and safely settle and adjust to life in a new environment are the causes of psychological disturbances and mental ailments affecting refugees. The other burning issues are essentially those of enjoying the legal rights stipulated in the UN Refugee Convention (1951) and the OAU Refugee Convention (1969).

Many refugees have died in conflicts, epidemics and personal attacks. Many have suffered physical hardships, and felt unwanted in human societies.

Some professionals and skilled refugees have become frustrated in seeking job opportunities. The elderly and disabled have experienced isolation and deteriorated, while the women and children are grossly abused. Refugees have been imprisoned, tortured as spies, and on occasion thrown back into their country of origin to face their adversaries.

Genuine attempts have been made by African governments, the international community and voluntary organizations to ameliorate the sufferings of refugees in every conceivable way. Much has been achieved, but invariably new problems arise in the course of implementation of even the most laudable programmes. It is the contention that policies governing refugee affairs in countries of asylum are a means of improving the image and the morale of refugees, and of preparing them to face yet another challenge of readjustment to life in the country of origin when they return.

Notes

1   Obtained from files in the Protection Division, UNHCR.
2   Facts obtained from files of UNHCR Headquarters, Geneva, 1982.
3   OAU Convention (Article III (1) and (2) – Prohibition of Subversive Activities).
4   Annual reports of UNHCR branch offices in Mozambique and Zambia, 1979.
5   UNHCR Branch Office, Mozambique, Annual Report 1980.

6   Information obtained from writer's interview with a refugee.
7   United Nations, Convention Relating to the Status of Refugees, Geneva, 28 July 1951,
    Article IC (1) –(6).

4 Solving the Problems of Refugees in Africa

Introduction

In a bid to find solutions to human problems the basic and the priority issue is survival. But humans also need a sound mind in a sound body to cope with the normal pressures of life, which for the refugee are real crises. Refugees need independence from the dictates of others and a measure of control over their lives and destiny, Since man does not live in isolation, refugees need recognition as part of any society in which they find themselves, being socially acceptable and able to harmonize their relationship with the human and material world. Refugees need assistance to achieve an acceptable standard of living and gain their composure and self-esteem, but more importantly, they need citizenship and all the benefits it offers. The success of this depends on positive attitudes and cooperation with those who rally behind them.

Refugee populations in Africa are enormous compared with the populations of the countries of asylum, as the majority of refugees are from rural communities. This factor, coupled with continuous large influxes, undermines the extent of the measures of assistance. Prolonged periods in refuge and the need for prolonged periods of assistance also worsen the depressed economies of the states so that enjoyment of the socioeconomic rights of refugees becomes limited.

The emotional and psychological problems of refugees can be reasonably handled if: their personal safety can be guaranteed; the countries of asylum are politically stable; and they are granted the fundamental freedoms to enjoy their legal rights under the Refugee Conventions.

Legal instruments which form the bases for solving the problems of refugees in Africa are, in the main, the Convention Relating to the Status of Refugees (1951) and the OAU Convention Governing the Specific Aspects of Refugee Problems in Africa (1969).

The UN Refugee Convention establishes three standards of treatment to be accorded refugees by a Contracting State:

1   'the same treatment as is accorded to nationals' of the Contracting State;

2   'the most favourable treatment accorded to nationals of a foreign country';
     and

3   'treatment as favourable as possible and in any event not less favourable
     than that accorded to aliens generally in the same circumstances'.

The first pertains to: freedom of religion, and the religious education of refugee children; free access to courts, such as legal exemption and exemption from *cautio judicatum solvi*; protection of industrial property such as inventions, designs or models, trademarks and trade names; rights in literary, artistic and scientific works; rationing; elementary education; public relief and assistance; and labour legislation and social security. These are subject to certain qualifications, fiscal charges, and wage earning employment, provided that refugees have completed three years of residence in the country, or have a spouse, or have one or more children any of whom possesses the nationality of the country of residence.

The second concerns the right of refugees to join nonpolitical and non-profit-making associations and trade unions.

The third applies to the acquisition of property and to leases and other contracts relating to moveable and immovable property; the right to engage on their own account in agriculture, industry, handicrafts and commerce; and to establish commercial and industrial companies. They have the right to practise a liberal profession, to obtain housing in so far as it is controlled by laws and regulations; and to obtain higher education.

The Convention provides that when the exercise of a right would normally require the assistance of authorities of a foreign country to whom a refugee cannot have recourse, the Contracting State should arrange for such assistance to be made available to him by government authorities, or by an international authority.

As a protection measure, the Convention requires a Contracting State to issue identity papers and travel documents to refugees lawfully staying in the territory of the Contracting States, unless there are compelling reasons of national security or public order.

Refugees lawfully staying in the territory of a Contracting State are accorded general freedom of movement. This includes the right to choose their place of residence and the right to nove freely within the territory, subject only to the regulations applicable to aliens generally in the same circumstances.

The UN Refugee Convention requires that:

> The Contracting States shall as far as possible facilitate the assimilation and naturalization of refugees. They shall in particular make every effort to expedite naturalization proceedings and to reduce as far as possible the charges and costs of such proceedings.

The OAU Refugee Convention also requires that 'Member States shall undertake to apply the provisions of the Convention to all refugees without discrimination as to race, religion, nationality, membership of a particular social group or political opinion'. The UN Refugee Convention gives support to this requirement.

These legal instruments provide the necessary guidelines for the development of national refugee policies in the countries of asylum that govern the daily lives of refugees. The countries are also guided by the spirit of the Declaration of Human Rights (1948) which gives the refugee the fundamental freedom of movement and association with nationals.

Some states are of the opinion that facilitating the enjoyment of refugee rights might work against voluntary repatriation. Consequently, curtailing of rights is adopted as a deterrent, one effective measure being to make it costly for the refugee to obtain the right. In some countries it is government policy to charge heavily for work permits and or naturalization applications to the point where refugees, already financially handicapped, are unable to meet the cost involved.

The office of the UNHCR is specifically called upon, pursuant to paragraph 8(a) of its Statute, to supervise the application of International Conventions for the protection of refugees. UNHCR's duty in this regard entails working closely with Contracting States so that national legislation and administrative policies are implemented and are consistent with the commitments that have been entered into at the international level. The UN Refugee Convention requires Contracting States to facilitate UNHCR's supervisory duty in relation to the Convention.

The UN Refugee Convention also requires that states undertake to provide UNHCR with information and statistical data concerning the implementation of the Convention, laws, regulations and decrees relating to refugees. The Protocol (1967) imposes the same obligations on states. UNHCR makes requests for specific information, on an ad hoc basis, on particular issues such as those granting Refugee Status, admission of refugees and their protection.

This information facilitates UNHCR's task in monitoring the practical application of the Convention. It also strengthens cooperation with the states in decision-making, disbursement of funds allocated to the country by UNHCR and in co-ordinating relief aid from other countries. Within the frame-work of annual reporting exercises, particularly in the area of protection, UNHCR field officers seek and continue to receive information from the states in preparing segments of annual reports to the Executive Committee and to the UN General Assembly through the Economic and Social Council.

The UN Refugee Convention states that:

> The Contracting States shall undertake to cooperate with the office of the UNHCR, or any other agency of the UN which succeeds it, in exercise of its functions and shall in particular facilitate its duty of supervising the application of the provisions of this Convention.

The Protocol (1967) confirms the need for Contracting States 'to cooperate with the office of UNHCR or any other agency of the United Nations', and the OAU Convention also requests Member States to cooperate with the UNHCR.

It is encouraging that in finding suitable solutions to the problems of refugees in Africa, not only African governments have to face the challenges; they have the active support of the international community.

More encouraging still is the interest being shown by other governments and organizations throughout the world. Five principal participants in refugee-assistance programmes in Africa are: the African governments – the countries of asylum; the United Nations and its agencies; the Organization of African Unity; some non-African governments; a large number of voluntary organizations – both national and international – including the churches. The refugees offer active support and participation in all programmes of assistance.

The basic elements needing to be addressed by national refugee policies are:

1   problems of reception at national frontiers;

2   liberalization of granting of Refugee Status, through established eligibility procedures;

3   humane treatment of refugees and asylum-seekers while in custody;

4   protection against refoulement, arbitrary detention and jail terms, and forced conscription into the army;

5 effective security against kidnapping, military attacks, air bombardments, etc;

6 respect for the Convention rights of refugees, and strengthening of the refugee family for national stability;

7 keeping refugees meaningfully engaged with income-generating activities;

8 promotion of the emotional and psychological needs of refugees;

9 special care of vulnerable groups: women, children, disabled, elderly and the sick; and

10 curtailing prolonged stay in refuge by encouraging refugees to return home when conditions improve, to resettle, or to obtain other citizenship.

Refugee Protection Measures

Protection of refugees covers a wide range of issues and is required in virtually all situations. Physical protection is against exploitation and forceful involvement of refugees in illegal activities, maltreatment during arrest for offences committed, assault by nationals and refugees alike, military incursions over refugee settlements, and assassination attempts. Furthermore, it is against emotional and psychological stress which refugees develop when in long periods of detention for having committed disciplinary offences. It is also against torture while in detention or imprisonment.

International protection means treating refugees according to international standards. Protection is against loss of access to eligibility procedures for refugee status; refusal by authorities to recognize applicants or to issue identity documents where these guarantee asylum-seekers the right to remain in the country during the period of screening, and the issue of travel documents with a Return Clause which afford refugees the legal right to cross international borders. More importantly, protection is against refoulement, deportation and extradition orders where refugees are forced out of the country of asylum. Protection for those living in restricted settlements where refugees face physical abuse from fellow refugees or from settlement authorities or from secret agents is a joint effort of the national security service and the refugees themselves.

Legal assistance is given in asylum procedures for granting Refugee Status; and for issuing identity and travel documents. Assistance may be limited to qualified legal advice, or to representation by a lawyer.

Good rapport between refugees and their supervisors is invaluable as a protection measure. Effort is made for ease of communication between refugees and authorities, as this is the medium through which refugees receive information on their obligations to the country of asylum, the rules and regulations governing life in refuge, and how their convention rights are designed for implementation. Information given freely as to developments in the refugees' country of origin is a means of protection as it enables them to plan their future.

Extensive information campaigns about the plight of refugees and their rights and responsibilities have been undertaken in the interest of refugees. This is in the hope that people all over the world will gain a better understanding of the circumstances and needs of refugees and give sympathetic support to their cause. On special occasions such as the 'Africa Refugee Day' every effort is made to inform the world about measures being taken to assist governments to receive and protect refugees, and to appeal for funds to meet the financial commitments to refugee programmes. An area of much concern is the appeal to all countries on this occasion to endeavour to maintain stability in order to stop the uprootment of their nationals.

Training sessions are regularly organized for the benefit of field workers including government immigration and border officials, and also non-governmental organizations working at international, regional and national levels, in the form of seminars, courses and conferences. The purpose is: to sensitize them to the international and regional Refugee Conventions and promote effective dialogue; to disseminate knowledge of national laws and regulations governing the protection of refugees; and to provide skills in handling refugees, particularly in human rights issues.

In many cases, the judiciary has an important role to play in protecting the rights of refugees against expulsion. Where the courts adopt an unduly restrictive interpretation of the provisions of the UN Convention (1951) and the Protocol (1967), such as 'persecution', this serves as a serious impediment to the full and proper recognition of the refugee. However, the ambiguous phraseology of the Convention itself allows considerable latitude for restrictive interpretation. Hence in certain national jurisdictions, the courts have differed over the proper interpretation of important provisions of the Convention.

The best form of protection given to refugees is for staff of the international community to be with them and serve as an international humanitarian witness in times of adversity. Usually junior field workers or civil servants in remote rural areas hesitate before committing an offence against refugees in the presence of staff of the international community. Countries are keen not to tarnish their image and colour world opinion on their deficiencies in protecting refugees. This is why it is necessary to have a staff of the international community at reception centres, during refugee relocation exercises, and in the settlements.

Between 24 and 28 August 1987, over 40 people from different parts of the continent came together in Maseru, Lesotho, to attend Africa's first conference devoted solely to the question of human rights documentation. Participants at the meeting took the concept of human rights in its broadest sense to encompass civil, political, social, economic and cultural rights, as well as collective and individual rights. Special emphasis was laid on the rights and needs of vulnerable groups such as children, women and refugees. The need for information exchange and a coordinated network were endorsed.

The official inauguration of the headquarters of the African Commission on Human and People's Rights was held on 12 June 1989 in Banjul, the Gambia. The Commission serves as the continent's coordinating point for promoting and protecting human rights based on the African Charter of Human and People's Rights.

Areas of particular concern are: promoting early warning of refugee movements; monitoring the condition of returnees; and ensuring that human rights information provided by refugees is used effectively to stem abuses in their countries of origin.

Refugee Assistance Measures

Survival of refugees and their satisfactory adjustment to life in the new environment is based on: assurances of their subsistence needs; access to health, education and employment; and enjoyment of their fundamental freedoms, including the ability to integrate into the community of the country of asylum.

Emergency Aid

Relief measures are directed essentially for a large influx of rural dwellers. They are employed at the time of reception of refugees to keep them alive at all costs while arrangements are being made for their settlement in the country of asylum.

In this effort, a large number of voluntary organizations local and foreign – as well as the staffs of the social services departments of the country of asylum– give active participation. Funds and goods are provided initially by the government, but if the burden is heavy because of the size of the influx, the international community is called upon to assist.

The challenges of relief aid pertain to:

a) the willingness of the government to shoulder the burden and grant asylum readily to the masses;

b) its state of readiness to open the borders as the people arrive;

c) security at the border to repel any attacks on the refugees;

d) the preparedness of relief aid workers to receive refugees with compassion;

e) adequacy of the relief aid package for the anticipated size of the influx – food, water, shelter, and effective medical aid, including personnel assigned specifically to the care of children, the aged and other vulnerable groups;

f) readiness of immigration authorities to carry out screening exercises promptly;

g) accessibility to other refugees wherever they have arrived along the borders – the state of the roads, availability of transport, ambulances and security measures;

h) moderation in the distribution of donated goods to refugees, not overlooking similar needs of their immediate hosts – those national communities living in the vicinity of entry points who are inevitably the first to show solidarity and offer hospitality. This effort, though it has the tendency to inflate the aid package, is an endeavour initially to portray the good image of the country as a whole, and also to avoid friction between the two groups;

i) preparation of a suitable site a distance away from the border where refugees can be settled; and

j) proper coordination of the entire programme of assistance.

A refugee census is always necessary in order to determine strategies for settling the refugees permanently as soon as possible. Many methods are used to get an estimated figure, such as the number of tents or other forms of shelter multiplied by an average number of persons staying in one shelter.

Sometimes the registers are used, counting by the number of families registered, multiplied by an average number of persons in each family. The government arrives at the number and gets this published nationwide to alert nationals and solicit their cooperation.

Sometimes the number of refugee population at the border is deliberately reduced to avoid national panic, or to avoid giving global impression of the level of destabilization in the country of origin. This is considered desirable if there is harmony between the two neighbouring countries. On the contrary, the government may inflate the number to worsen bad relations with its neighbour; or do so with the expectation that a great deal of relief aid would be brought to the country for the benefit of both refugees and nationals.

It is important to limit the period of relief aid to a minimum, relocate the refugees expeditiously to avoid border clashes, and to settle them to a normal life. This effort requires the active participation of the refugees, their initiative and cooperation in measures being planned for their future. Operational partners are usually appointed to deliver specific services at the initial stage and to continue their services at the settlement sites.

Providing food, shelter, clothing and medical care for large numbers of refugees is a costly and complicated exercise. Donors have to be found locally and abroad; and goods have to be transported, thus raising the matter of transport, fuel and storage facilities. Manpower for relief aid, often drawn from national resources, disrupts services to nationals. Relief aid personnel coming from abroad have to be sponsored by the voluntary agencies, but government has to guarantee their security in the country. The entry of a large number of international observers and the press poses problems of accommodation and transportation, and it also causes a strain on the social services of the country.

Within the set-up is the active participation of national volunteers and social workers assisting families with small children and other vulnerable individuals or groups. The Red Cross and the Red Crescent identify the sick and injured and convey them to health posts, while volunteers organize children's programmes to boost their morale. Religious organizations also seek opportunity for worship with refugees and administer any religious rites considered necessary. Generally, a programme of social activities that involves as many of the refugees as possible has been beneficial.

At the end of emergency aid, refugees now settle to a new life and take advantage of socioeconomic assistance of the country of asylum. A number of stakeholders involved in this exercise are: the government, UNHCR and other UN agencies; UNHCR operational partners, and voluntary organizations, including the churches. The main objective is to work towards reverting refugees to their original status as citizens of a country, having maintained their physical and mental faculties to continue living a normal life.

The Role of Governments

In respect for international and regional Refugee Conventions, and the overwhelming support received from the international community and the nation as a whole, governments have made invaluable contribution to the welfare of refugees and continue to do so. They give humanitarian consideration to the reception of asylum-seekers and grant them a legal status and some socioeconomic rights. Governments provide vast areas of land in remote districts to build settlements for refugees and to make it possible for them to cultivate it. They allow a host of personnel of different nationalities from abroad to enter the country in the field of refugee assistance. They facilitate the importation of goods and services for refugees and allow refugees to utilize the social services established for nationals. Above all, governments give as much protection as is reasonably possible to refugees within their territories.

Granting refugee status to many thousands of people within a short space of time could create serious economic and social constraints on the country of asylum. This is particularly the case for poor states where the willingness of nationals and the inclination of the government to shoulder the burden is greatly diminished.

Under such circumstances there is conflict of responsibility between international obligations and national responsibilities. Consequently, in a number of states, priority is accorded to nationals over all aliens, including refugees, in fields such as education, employment and housing. Even under normal circumstances any Contracting State can be expected to face some domestic opposition in securing the basic needs of a particular group of aliens.

Malawi, a tiny country of eight million people, gave refuge to over half a million Mozambicans. Burundi is one of Africa's smallest countries, with a population not exceeding 4.6 million in 1978, but in spite of her enormous economic problem, she continued to bear the brunt of hosting some 198,000 refugees – 65,000 Rwandans, 9,650 Zairians, 930 Ugandans and urban refugees of various nationalities – according to UNHCR estimates. The government, however, adopted restrictive measures with respect to jobs and access to education and health services. The reasons for this new attitude were the saturation of the labour market and the desire to protect the country's own nationals in a difficult economic situation.

A great deal of effort has been made by governments to educate the general public on refugee matters through the mass media. Thus refugee integration into communities has been enhanced through understanding of their situation, and has also encouraged local communities to share resources in a humanitarian spirit. The governments of Botswana, Tanzania, Zaire and Rwanda, for example, have encouraged assimilation of refugees into the national community.

A substantial number of refugee students have been accepted for studies in almost all universities in Africa: Egypt, Nigeria, Ivory Coast, Sierra Leone, Ghana, Gambia, Congo, Burundi, Kenya, Zaire, Ethiopia, Zambia and others. Students are granted temporary asylum during their period of study.

As an incentive to refugee children in educational institutions in the Sudan, the President of the Sudan attended the graduation ceremony of nine refugees who won awards from the Piastre Institute in Khartoum.

The Role of UNHCR

In pursuance of the provision of the UNHCR statute the office has, since 1961, planned and implemented a wide variety of programmes in Africa for legal protection and material assistance to refugees. UNHCR presence is felt in all the countries of asylum no matter what the size of the refugee problem.

Subsistence needs are provided for individual refugees. Rural settlement schemes and self-sufficiency projects for refugees are given the necessary support. Social infrastructure of government in health and education are expanded, and if unavailable are established for the benefit of both refugees and nationals.

UNHCR establishes self-sufficiency projects and, when these become viable ventures, hands them over to governments. The Mishamo settlement for 34,000 Burundi refugees in 16 village communities run by UNHCR and LWF is one such example which was started in 1978 in the vast unpopulated tsetse fly-infested Miombo wilderness of western Tanzania, and handed over to the Tanzania government in August 1985.

Together with the governments concerned, UNHCR coordinates all protection and assistance to refugees. More importantly, in times of adversity, UNHCR keeps vigil on behalf of refugees, and is physically present to give moral support and assistance. UNHCR also gives its cooperation and directs moral persuasions to governments to grant refugees a legal status upon which their legal rights for protection and material assistance are based. In support of these programmes UNHCR appeals for funds and finances its projects with millions and millions of dollars in Africa.

Operational Partners

Projects for refugees are undertaken by operational partners appointed by the UNHCR and the governments of the countries of asylum. A prerequisite for any implementing partner is that the partner must be willing to work with all intended beneficiaries, regardless of their race, religion, nationality, political opinion or gender; and provide assistance on the basis of agreed needs only, without linking this, either directly or indirectly, to any ethnic, religious or political consideration.

Furthermore, they must be: legally registered at the location of their headquarters and/or in the country where they are operating; have the authority to operate a bank account in the country of implementation and the ability to maintain separate accounts for any expenditures incurred; be able to demonstrate financial reliability through the production of official audit statements; and establish a working relationship with the government that allows them to operate within the country concerned. The partners should be able to demonstrate ability to deliver assistance effectively.

The major partners are: the World Food Programme (WFP), a UN agency in Rome, Italy, which is the food-aid arm of the United Nations system; the World Council of Churches (WCC); Lutheran World Federation (LWF); and OXFAM. Operational partners are guided by the necessary operational policies of government. Where necessary or feasible, government agencies and local private corporations have also accepted responsibility for service delivery to refugee settlements.

Voluntary Organizations

According to the United Nations High Commissioner for Refugees,

> Non-governmental agencies (NGOs) are vital partners of UNHCR in the field –
> in countries of asylum and in donor countries, in the implementation of
> programmes, raising funds, persuading governments and sensitizing public
> opinion. I greatly value their role and fully intend to strengthen UNHCR
> collaboration with them.
>   At the same time, one must take note of the rather astonishing proliferation of
> NGOs in the refugee field (*Refugees*, No. 33, November 1986, p. 19)

The voluntary agencies are entirely indispensable as implementing
agencies for rural development projects. They do not only supervise on-the-
spot day-to-day activities in the field, but they are also a source of great support
in the social adjustment of refugees. They create incentives for the local
communities which serve as a means for reinforcing good relations between
refugees and nationals.

In order to strengthen refugee families voluntary agencies help trace the
whereabouts of members of separated families when correct information on
the mode of separation is known. When family members are located, all efforts
are made – financially and through contacts with other agencies – to put them
together in one country of asylum if they so wish. If reunion is taking place in
the same country of asylum, transportation is paid for. If a member of the
family has to travel to a neighbouring country to join the others the procedure
is complicated by the acquisition of travel documents and visas. This is time-
consuming and involves much financial support.

The guiding principles influencing all decisions concerning
unaccompanied minors in the care of other families are in the best interests of
the child for normal family life. Legal responsibilities for all such decisions
formally rest with the authorities of the countries of asylum. The organizations
exercise certain protection responsibilities by obtaining an assessment of their
immediate needs, an analysis of the long term and immediate viability and
foster arrangements.

Other Participants

Regional organizations such as the OAU and UN Economic Commission for
Africa (ECA) have also achieved much in helping refugees in Africa. The

OAU project for placement and education of African refugees and its scholarship awards were laudable. The ECA carries out studies and lends logistic information towards the assistance of refugees.

Settlement Schemes

Settlement schemes in the country of asylum, planned and successfully implemented, are the bases upon which the great majority of refugees are able to adjust to normal living conditions – live as families, prepare their own meals, buy their own clothes, make their own transport arrangements, practise their religion, and adhere to their cultural and traditional values.

Four types of settlements are planned for refugees: urban settlement; assisted spontaneous rural settlement; non-assisted spontaneous rural settlement; and organized rural settlement.

Urban Settlement

This type of settlement serves the interest of refugees originating from cities and towns such as administrators, politicians, members of the learned professions, security personnel and students. It promotes the refugee's efforts to live privately and manage his own affairs. It encourages easy integration into the community and affords opportunity for seeking avenues for social advancement. Marriages are contracted, thus paving the way for assimilation and to naturalization in the country of asylum.

General assistance Urban refugees are sponsored by the NGOs. The organizations give welfare assistance to those refugees who have no other means of support. Each person is dealt with on an individual basis.

The organizations do not recognize any committee of refugees or any group of refugees as being representative of any or all refugee interests. Any and all assistance given is an emergency measure and not a long-term solution.

If a refugee has once been locally employed or maintained by other self-help measures and he subsequently loses that opportunity, he is not automatically entitled to assistance again. Refugees who return or are returned by another country to the first country of asylum, after having left it, are also not automatically entitled to be assisted. The maximum period of assistance provided by the organizations is two years. This does not mean that every refugee is assisted for the maximum period of time because each refugee is treated on an individual basis.

It is a policy of the organizations that any and all assistance to refugees is not higher than that of an average citizen or family of the region where the refugees reside. Destitute refugees are given 'settling-in' allowances and monthly subsistence allowances.

Primary school allowances are, however, provided for dependants during the period of assistance to the refugee. No allowances are provided for secondary school expenses. In cases where dependants of a refugee need educational funds, parents are referred to agencies for an award of scholarship or bursary.

The provision of vocational training has also been an invaluable part of assistance. Scholarships are awarded for study and training in agricultural and vocational institutes which cover the supply of equipment and books. Awards of grants-in-aid are made to existing schools to enable them admit refugees.

In the urban settlement scheme, refugee students, for instance, live among other aliens and nationals in the same institutions. Assistance with food, accommodation, clothing and transport for refugees can easily be compared with the economic conditions of their colleagues. Any disparities and any special assistance in favour of refugees cause friction between them and the others. In some countries, failure on the part of assisting agencies to detect such anomalies has resulted in the refugee students being assaulted and even to a national outcry against all refugees.

Some refugees in urban settlement have suffered hardships of inadequate allowances to meet the rising cost of living. Consequently, refugees compromise on food, shelter and clothing and ruin their health. They are evicted from rented quarters for arrears in rent. They are deprived of membership of societies that may otherwise offer them some social satisfaction, because they are unable to pay membership dues.

Refugees are subjected to all kinds of discrimination at work, at medical centres, at educational institutions and at their places of residence. These problems are not conducive to normal adjustment efforts. They make refugees disoriented and homesick.

Counselling of refugees Counselling services have been established in almost all countries of asylum by UNHCR and the voluntary agencies, especially the churches. These have proved of increasing importance and value in the measures being taken to integrate them into the social and economic life of national communities. The scope of the service is very broad. Essentially, it gives refugees orientation to living conditions, to the social services available to them, and to channels of communication. It undertakes the administration of subsistence and supplementary allowances to needy refugees.

Special assistance is given to refugees needing educational placement locally and abroad, particularly scholarship students whose awards are channelled through the counselling service. Refugees having special health problems receive financial assistance, moral support and guidance. The service lends support to those refugees who need loans to establish economic ventures, which may ultimately help them to achieve self-sufficiency. When the need arises for travel the counselling service negotiates with the host government, on behalf of refugees, to hasten the procedures for obtaining travel documents. The counselling service also helps to facilitate repatriation and naturalization for those refugees who decide to revert their status and become nationals of a country of their choice. The refugee counsellor is a trained social worker experienced in the care of vulnerable groups in society.

He is well-versed in refugee law and the rules and regulations governing refugees in the country of asylum, and often gains access to the Refugee Commissioner to acquaint himself with economic and social developments in the country. The refugees look up to the counsellor for advice and assistance in their day-to-day activities. In order to ensure proper supervision of refugees, all refugees in urban areas and areas of counselling operations are registered with the counselling service.

In the early 1970s, with increasing numbers of individual refugees in capital cities of countries of asylum, there was the need to intensify efforts of supervision of refugees. A professional social worker, an African, was therefore appointed by UNHCR to undertake this assignment.

He travelled from one African capital to another to advise on the establishment of new counselling services and to recommend ways of upgrading or restructuring existing ones. The new approach is reflected in the counselling service established in Bujumbura, Burundi, in June 1974 (UNHCR Annual Report, 1974).

The scope of the service has further been broadened to cover various types of assistance projects to individual refugees such as welfare services for the aged, infirm and the handicapped, and various forms of loan schemes for housing and private business enterprises.

The refugee counsellor's position is not an enviable one because of the nature of assistance he is required to give, and the rules and regulations governing each kind of service, especially where these involve the allocation of subsistence allowances and job placements. This is of particular reference to refugees who refuse to work when work is obtained for them, or to those who expect a lion's share of allowances.

Counselling has been very beneficial to refugees, as an essential component of refugee assistance, and should be a special feature in national refugee policies.

Scholarship students Generally, educational policies require that students applying for assistance present a written statement from the minister responsible for Refugee Affairs confirming that they are refugees in order to be considered. Eligibility for admission to the programme is by passing an entrance test. Each student admitted to the programme is issued with an identity card which gives him the right to receive monthly food and lodging allowances, school supplies, and annual allowance for bedding and clothes. The office does not replace any lost or stolen items such as household items and clothes.

Students are given free medical assistance but must obtain coupons in order to get medical treatment. Allowances are temporarily withheld during a period of hospitalization and promptly resumed upon discharge. The office refunds payments for medications only if they are prescribed by a physician. Eye glasses are paid for by the office only when prescribed by an optometrist or optician appointed by the office. The office pays only for tooth extraction, as regards dental care. Filling and replacement are paid for only when found medically necessary.

There are policies governing offensive behaviour, theft and misappropriation or wilful neglect of the property of an educational institution, intoxication with alcohol or substances of abuse and continued failure to pay for food and rent charges.

An annual evaluation is made of the student's academic performance, and if he fails for two consecutive years or has derogatory reports on conduct, he is expelled from the programme.

Students threatened with expulsion from the programme have a right of hearing by a disciplinary committee and can appeal against adverse decision to the minister responsible for Refugee Affairs. Any student who marries after admission to the programme is not entitled to family assistance of any kind; he continues to receive the allowance afforded a single person. This programme is typical in Ethiopia.

Some refugees have indeed benefited from these educational schemes, but their number is small while the type of education and training given so far is far from the standpoint of the economic and manpower needs of Africa.

Employment The employment of refugees requires that refugees be issued with work permits. The minister responsible for Refugees Affairs, at his discretion, instructs the renewal of such work permits for such periods as he deems fit. The validity of such work permits may be extended for more than 12 months. If such provisions are not made, refugees who have no other country to go to will have to remain in the country of asylum and receive public assistance indefinitely. It is considered that where a refugee has a special skill such as teaching or nursing to contribute to the government he should be permitted to work provided there is no such citizen available for the job.

Assisted Spontaneous Rural Settlement

Under this scheme, skilled and semiskilled refugees with or without ethnic affiliations with the rural communities, are self-settled among the local population. In order to obtain the necessary assistance, they are registered with refugee-assisting agencies operating in the locality. Food rations are obtained from the WFP. Access to community social resources is secured through international assistance measures, where already-existing facilities – health institutions and schools – are extended for the benefit of refugees. They are encouraged to accept employment offered in the private sector in urban centres or in rural enterprises. In general, subsistence allowances are withheld where self-reliance possibilities exist. Refugees settled in this way apply for access to land in competition with local inhabitants.

Like urban refugees, spontaneous settlement affords refugees the opportunity to integrate into the society. Tens of thousands of refugees in assisted spontaneous settlement in the Sudan, for instance, have migrated to the towns in need of greener pastures.

They are estimated to number some 200,000, with 100,000 in Khartoum alone, and account for up to a quarter of the national population in certain towns in the east. The consequences are obvious – an outcry by nationals against the presence of refugees.

Non-assisted Spontaneous Rural Settlement

Refugees live scattered among a given national population in rural areas. They have ethnic ties with the local national community, speak the language and dress like them, so that it is difficult to tell them apart. They exist independently more or less on the same basis as rural migrant workers with no official recognition, or may operate in self-sufficiency ventures.

They provide all their personal needs without resort to any refugee-assisting agency and register with them for assistance only when they become destitute. Non-assisted spontaneous rural settlement has taken place in border areas in almost all major countries of asylum including Uganda, Ethiopia and the Sudan.

Success with self-sufficiency efforts promotes local integration and eventual assimilation into the community. However, when the number of such refugees becomes visibly on the increase and food prices increase housing becomes inadequate and rent increases; unemployment rate rises; and pressure on land, fuel and water supply reaches unbearable limits, the feeling of xenophobia takes hold of nationals. Politicians put the blame squarely on refugees, thus generating tension between refugees and nationals.

Organized Rural Settlement

Large rural refugee groups transferred from emergency relief centres around the national frontiers settle in ethnic groups; or occupational groups in food crop farming or in fishing; or as craftsmen or as herdsmen. The settlement is usually located in a sparsely populated district designed by government for refugees and laid out as protected villages.

Problems arise where there is a massive influx of new refugees who cannot be absorbed on the basis of ethnic ties or occupational skills. Where the local district into which refugees are being received has residents of different ethnic groups, utmost attention needs to be paid to the local sociopolitical structure, particularly if it is being considered as a possible permanent settlement area.

To explain the social processes involved and to suggest positive ways in which an impasse between the refugees and the administering authorities can be avoided, there is the need to consider the importance of carrying out, in stages, the development of the settlement and the internal organization of refugee communities. Once a refugee community is established, its members almost inevitably determine their developmental and adjustment strategies.

Closely connected with land settlement has been the drawing up and implementation of integrated regional or zonal development plans in the country of asylum. Some authorities regard this as the third and final phase of the settlement schemes. The main purposes are twofold. The first is to render the refugee settlements not merely self-sufficient but also economically and socially viable, and to enable the refugees earn income through economic and social activities.

The second is to develop the agricultural, social and infrastructural facilities of the settlements on a more permanent basis and extend them in such a way as to benefit the local population, as well as to integrate the refugees. In other words, the refugees are to be converted into active elements of economic and social development. Their presence in the various countries is thereby transformed, as Saddrudin Aga Khan puts it, from 'a liability into assets of the economic and social balance of the countries' development'.

These development plans have included projects calculated to raise the living standards of the refugees and that of the general rural population, through the transformation of prevailing subsistence economy into a monetary one; to increase their incomes and improve conditions of life through the construction of schools, workshops, dispensaries and play grounds; and to provide better housing and other facilities. Particular attention has also been paid to the production and sale of protein-rich foodstuffs and to the supply, where possible, through cooperative organizations, of consumer goods at reasonable prices in these areas. The development of agricultural facilities based on the production of cash crops, such as tea, coffee, groundnuts and cashew trees; and the encouragement of the development of livestock activities on commercial basis, have been further aspects of the regional development programme.

Refugees as an Asset

Many argue that the presence of refugees has generated international financial support for national development programmes. However, refugees have not always been passive recipients of developmental aid. They have been creative and industrious, and have posed a challenge to host nationals to upgrade their standards of achievement. Some refugees have even lent their expertise in various fields to the service of the country which offered them asylum, thus setting an example of self-sacrifice for nationals to emulate.

Refugee youths, through hard work and determination, have won scholarships to study abroad. Some have returned to put their knowledge and experience to the service of Africa. A refugee child won the first prize for the best student of the year at Qala-en-Nahal in the Sudan in 1973, thus paving the way for the enrolment of more refugee boys at the school. In 1984, 40 refugee students from Rwanda finished their studies in medicine, architecture, economics and law at the University of Dakar, Senegal.

The construction of a community centre in Dein El Nair, Port Sudan took two years of planning and six months of building by refugee volunteers. The

Ndzevane agricultural settlement in Swaziland for 10,000 South African refugees is profitably engaged in cotton cultivation, the mainstay of the settlement. Self-help projects which have earned admiration for refugees include poultry and pig raising; and duck, turkey and rabbit farming.

A Sudan refugee settlement consisting of 10,000 women has successfully organized a life without men. At the border between Sudan and Chad almost all the men died in the Chadian civil war. Fourteen women were elected by the settlement's seven ethnic groups to run its massive feeding programme. They took care of 4,000 children.

After many years, some 20,000 south-Sudanese in the Central African Republic have returned to their country and a refugee problem has been solved in the best possible way. The Central Africans, however, regretted their departure because the refugees brought economic and social development to the remote and sparsely populated southeastern corner of their country. These efforts affirm the view that refugees, when granted a congenial environment, can be an asset to society.

Repatriation

This is an opportunity for refugees to go back where they belong. The voluntary nature of repatriation is a Convention (1951) requirement. The refugee may decide to return to his country of origin when the conditions which caused his uprootment have ceased. He may otherwise decide to return home in spite of existing hostilities there, if he finds conditions in his country of asylum deplorable, especially if these conditions are worse than those in his country of origin. He may also decide to return home by taking advantage of an amnesty granted by his country of origin to persons in his group.

Promoting and facilitating voluntary repatriation of refugees is a fundamental protection function. In 1988, more than a quarter of a million refugees availed themselves of the opportunity to return voluntarily to their country of origin to re-establish themselves among their ethnic communities. The main movements involved the voluntary return of 80,000 Ugandans, 69,000 Mozambicans, 53,000 Burundians, 7,000 Ethiopians, over 3,000 Zimbabweans and about 1,000 Chadians (*Refugees*, Special Issue, December 1988, pp. 10–22).

The OAU Convention also requires that repatriation be on voluntary basis: that adequate arrangements be made for the safe return of refugees requesting repatriation; that the country of origin, on receiving back refugees, should grant them full rights and privileges of nationals; and that refugees who voluntarily return to their country should in no way be penalized for having left it.[1]

UNHCR works actively to promote respect for its guidelines on voluntary repatriation. One of its basic provisions is the right of refugees to return as well as to receive information on the prevailing situation in the country of origin so as to enable them reach a free decision concerning their return. It observes the voluntary and individually manifested character of repatriation; the necessity for repatriations to take place in conditions of security and dignity; and the recognition of the right of refugees to choose their destination in the country of origin.

Other provision are: the principle of non-recrimination for having sought asylum, together with continued respect for the cultural and ethnic values of the returning refugees; continued respect for access to means of subsistence and to land under the same conditions as other nationals of the country of origin; and respect for UNHCR's work in favour of returnees and the office's access to them.

These basic provisions were included in voluntary repatriation agreements concluded by UNHCR and governments to facilitate voluntary repatriation. They also provide for the comprehensive monitoring of the return movement. In anticipation of future voluntary returns, UNHCR has significantly strengthened its field staff with a view to being able to assess the voluntariness of return and to monitor the return journey.

Namibians have been assisted in this way. UNHCR established a large presence to assist in the logistics of the operation as well as to monitor the protection of the returnees. In order to facilitate a voluntary return in safety and dignity, UNHCR was instrumental in the promulgation of a blanket amnesty for the returnees, as well as the repeal of a large number of discriminatory laws. As the first voluntary repatriats arrived in Windhock on 12 June 1989, almost 40,000 candidates for repatriation had registered with the UNHCR abroad. An earlier repatriation exercise for over 300,000 Angolan refugees in Zaire was carried out in 1976 by UNHCR with the support of other United Nations agencies.

Voluntary repatriation programmes are, however, not without their share of problems. In some instances, pressure is exercised on refugees by governments to return to their country when conditions improve. In one instance, for example, the registration form the refugees were requested to complete did not clearly indicate the purpose for which it was used and it contained no declaration of voluntariness to return.

In some other instances applicants for voluntary repatriation programmes were put under pressure by other refugees or 'interest groups' not to return. On several occasions UNHCR was able to intervene to ensure continued respect for the individual's right to opt to return and to enjoy his safety and dignity, by separating the candidates from those opposing return.

Continuous influx of refugees often minimizes the effects of successful repatriation of large groups of refugees. In spite of efforts at repatriation out of the Sudan, there was a massive influx of refugees into the country – 17,000 Ethiopians entered eastern Sudan on 25 December 1984, rising to 58,000 on 9 January 1985, and at the same time between 60,000 and 100,000 Chadians entered western Sudan.

Where refugees are uprooted by double calamity – hostilities as well as drought and famine – they should normally repatriate when the hostilities which formed the basis upon which they were granted refugee status have ceased in the country of origin. But because of the drought and famine still prevailing in the country of origin refugees are unwilling to return home. This situation exposes the refugees to harassment and possible ejection, as they have now lost their refugee status. UNHCR intercedes on behalf of such refugees by pleading with the government to let them stay on humanitarian grounds. Nevertheless, not all of the office's interventions are met with success.

In order for voluntary repatriation to have a large impact on the overall refugee situation, much more needs to be done. Only where conditions which provoked the original refugee movement have disappeared can greater numbers of refugees be expected to choose voluntarily to return to their country of origin.

Naturalization

The refugee may choose to remain permanently in the country of asylum because, somehow, he has successfully established himself there. In this circumstance he reverts his status by naturalizing, that is, obtaining citizenship of the country of asylum. An asylum-seeker, on acquiring Refugee Status, relinquishes the protection given him by his country of origin, or of any other country where he previously claimed citizenship. Consequently, in order to obtain the protection of another country, the refugee needs to become a citizen or national of that country. If the refugee wishes to acquire citizenship of another country other than the country which has granted him asylum, he must in the first place resettle in a country of choice, then apply for citizenship.

In keeping with the UN Convention (1951)

> ... the Contracting States shall as far as possible facilitate the assimilation and naturalization of refugees. They shall in particular make every effort to expedite naturalization proceedings and to reduce as far as possible the charges and costs of such proceedings.

In order to hasten the day when a refugee, if he does not wish to repatriate, can become a citizen of another country, the UNHCR office encourages governments to make special concessions regarding the acquisition of nationality. The office also makes refugees aware of the opportunity for acquiring new citizenship which could earn them the rights and privileges of nationals, and advises them on the legal procedures involved. Success has been achieved in establishing a naturalization project for Angolan refugees in Botswana and for Rwandan refugees in Tanzania, in waiving the fees for naturalization of destitute refugees in Burundi, and in shortening the mandatory residence period in other asylum countries such as the Sudan.

In Somalia, the Refugee Commissioner announced that his government had since 1983 made it an official policy to offer residence to any refugee who could exhibit self-sufficiency.

Resettlement

This may be defined as the immigration of refugees from the country in which they have been granted asylum to another country which could offer them the possibility of permanent settlement. It has been found to be the most suitable solution for three principal categories of refugees all of whom have an urban background. The first category consists of artisans, white-collar workers, students and members of learned professions who cannot, for reasons of economy, be absorbed into jobs in their country of asylum. The second category comprises African refugee graduates from universities or higher technical institutions, either in Africa or abroad, who wish to place their newly acquired skills to the greatest possible advantage in Africa. The third includes refugee students who seek educational assistance either in the technical or general academic field in countries other than the country of asylum.

Resettlement also serves as a protection measure for individual refugees requiring urgent relocation to ensure their survival in safety and dignity. These include those refugees threatened with immediate deportation or refoulement but who, on account of humanitarian considerations, need not return to their country of origin; and those who have cause to fear assassination in the first country of asylum. It also protects refugees living in segregated rural settlements where prolonged stay in a confined area has led to a range of human rights abuses; and also to those persons being tortured while in prison or detention.

Resettlement can take place only if refugees can travel from the country of asylum to their new country of adoption. This raises a number of interrelated technical problems which can only be solved through the understanding of interested governments. In the first place transport costs, and for students, scholarships have to be provided. Secondly, the asylum countries have to issue travel documents with a Return Clause to refugees to facilitate their free movement and their acceptance. Unfortunately, certain governments have been reluctant in issuing return visas. Thirdly, the country of adoption should not only grant entry visas but should also make it possible for refugees to have some privileges – to receive education, training, or employment; and to designate an educational institute, voluntary agency or governmental organization, as appropriate, to assume responsibility for them. Furthermore, the country has to arrange for the reception of the refugees.

In practice also, governments have allowed refugees to resettle in second countries of asylum, which means that much of the favourable treatment required by the refugee Conventions are recognized by the Contracting States, without necessarily being included in the refugee laws.

Urgent resettlement for the protection of refugees has been a major exercise. Immediate and vigorous negotiations are taken up with the second country of asylum till some success is achieved. The government of the second country of asylum issues the necessary visas but travel documents are issued by the government of the first country of asylum. As often as the case demands, the refugee is escorted to the aircraft and is met at his destination.

International Measures

In the last decade, the impact of the world economic recession, combined with environmental disasters and population growth, has produced an unprecedented crisis, both social and economic, in many Third World countries. Sadly, but understandably, a number of these nations have already concluded that refugees represent an unacceptable strain on their limited resources.

Responding to this critical situation, members of the international community have in recent years developed a new concept of 'refugee aid and development' assistance programmes which are designed to produce self-sufficiency rather than dependence, which benefit refugees and the host community alike, and which make a positive contribution to the overall development of areas where refugees live.

Putting these principles into action has not been easy. At a time of financial stringency, many donor countries have adopted a cautious approach to new and potentially costly forms of aid. Host governments have been worried that if donors agree to provide development-oriented assistance for the benefit of refugees, then they might be tempted to pay for this by reducing their contributions to both general development and refugee assistance programmes.

Finding the right organizations to put the new concept of refugee aid and development into practice has also raised problems. The international community has hitherto tended to regard refugee aid and development as separate activities, with UNHCR responsible for the former 'aid', and other organizations such as the UN Development Programme (UNDP) and the World Bank, responsible for the latter – 'development'.

Many non-governmental organizations (NGOs) in recent years have been preoccupied with emergency situations in areas such as northeast and southern Africa, thus diminishing their role in the implementation of development-oriented refugee assistance programmes.

Despite these obstacles, the concept of refugee aid and development has now been placed firmly on the international community's agenda.

Moreover, the concept which has been carefully formulated in discussions between the government and organizations concerned, is now being put into practice and has brought visible benefits to refugees and host communities. In pursuit of this concept there is collaboration with the International Fund for Agricultural Development, whose traditional beneficiaries are the poorest people in the rural areas where refugees are frequently located. Another important effort to establish a practical link between refugee aid and development began with the convening of the second International Conference on Assistance to Refugees in Africa (ICARA II) in June 1984, at which 112 nations, many UN agencies and 145 NGOs were represented.

ICARA II has had two positive results. On one hand, in the words of author Robert Gorman, it 'sharpened and focused an already emerging dialogue on the connections between refugees and development'. The Conference's final declaration and Programme of Action was of particular importance. It confirmed that as a result of the diverse impact on the national economies of the African countries concerned ..., there is a need to provide these countries with the required assistance to strengthen their social and economic infrastructure, so as to enable them cope with the burden of dealing with large numbers of refugees and returnees. This assistance..., should be additional to and not at the expense of concerned countries' other development.

At a more practical level, ICARA II initially produced around US$ 55 million in pledges from donor countries for infrastructural projects in Africa, some of which are now at the implementation stage. Examples include a fish-farming project in Zambia, the construction of grain-storage facilities in Somalia, a cattle ranch in Rwanda, and a water supply programme in the Sudan. Despite these achievements, the concept of refugee aid and development is still at a formative stage. Although it is still the subject of debate and discussion, the concept is now central to efforts to provide durable solutions for the world's refugees, and should form part of national refugee policies.

Note

1   OAU Convention Article V.1–5.

5 Refugee Policies of African Countries

Introduction

The United Nations Convention (1951) makes it incumbent on Contracting States to recognize the Rights of Refugees in their legislations. Consequently, every endeavour is made by governments to legislate and develop national policies to direct implementation measures. Policies, when comprehensive and based on moral and humanitarian grounds, are presumed to afford refugees an opportunity to shape their lives to the best advantage, and to inculcate a sense of trust that the government has their interest at heart. The mere fact that a Minister of State is assigned legal responsibility for refugee matters is ample evidence of the impetus refugee matters have in the affairs of the countries.

Refugee policies have been put to the test in some major countries of asylum to ascertain the amount of impact they make on refugees. The fact has been established that some aspects of the policies are not appropriate for the management of the majority of refugees in both urban and rural settlements, and certainly not so supportive in forging harmonious relationships between refugees and the government, particularly where protection is concerned. The nature of some policies has given cause for concern in their formulation interpretation, even in the courts; and at their implementation stages.

Three national refugee policies in major countries of asylum in Africa have been selected for assessment. They have survived decades of apparent stability in the respective countries, while those in the other major countries of asylum have been influenced over the years by frequent change in government.

a) The Refugees (Control) Act, 1965 of the United Republic of Tanzania.

b) The Refugees (Recognition and Control) Act, 1967 of Botswana; and

c) The Refugees (Control) Act, 1970 of the Republic of Zambia.

The policies under review stress the significance and relevance of national refugee policies as direction, guidance and a measure of control are necessary in refugee affairs. Some countries do not have such statutory instruments as their involvement with refugees is not so significant, and others relate refugee matters to immigration laws.

Areas of Concern

Inadequacy of Policies

Some policies fall far short of expectations. The rights of refugees are not adequately covered. Subsistence, shelter, health and education are hardly mentioned. They appear collectively in statements where administrators are 'to ensure the performance of any work or duty necessary for the maintenance of essential services in any refugee settlement or for the general welfare of refugees therein'; 'to ensure that all proper precautions are taken to preserve the health and well-being of refugees therein'.

Indeed, policies without the necessary tools for their enforcement are not worth formulating. The countries may not have adequate machinery to enforce certain rights of refugees, hence, they are excluded in refugee policies. Where a Contracting State has a commitment to grant the rights of refugees, and where the international community gives overwhelming material and financial support to governments, it is pertinent that all the rights of refugees form essential ingredients of the policies. In their absence, the humane and favourable treatment of refugees is consequently lost and the good intentions of government are overshadowed by other measures considered to be in the best interest of administration.

Favourable refugee policies encourage refugees to contribute productively their talents and skills for the well-being of the country that has offered them refuge. Such worthwhile contributions also relieve the government of the burden of care.

Discretionary Powers

The management of refugee affairs by a high-ranking government official such as Minister of State is all-embracing, but the power vested in him to use his discretion in making rules and regulations makes room for arbitrary treatment of refugees. This is particularly serious in matters relating to Refugee Status and to the care of refugees in organized spontaneous rural settlements.

For instance, a minister may exploit his discretionary powers to recognize certain aliens as refugees on account of their political value to the government. This enables such persons to benefit from assistance programmes provided by voluntary organizations, particularly those awarding scholarships for higher education, while refugees lose such opportunity. It is important that the minister has advisors to assist him in taking decisions.

Policy Design

The ambiguous character of some policies is open to different ways of interpretation. A word that has caused much controversy as an eligibility criterion, even in the law courts, is 'persecution'. In the policies, expressions such as 'offences which prejudice peace and good order' and to the refugee 'acting in a manner prejudicial to peace and good order' are difficult to explain. They are not related to circumstances in which those offences are committed.

Not all refugees have the benefit of communication with refugee authorities. Refugees in non-assisted spontaneous rural settlements do not have contact with refugee counsellors. Hence, they are not aware of refugee obligations or the manner in which they are expected to conduct their lives. They are the ones who may flout these laws unknowingly. It is necessary that refugees, wherever they are in the country, have the means of communication with authorities and if possible hold occasional meetings with refugee counsellors to receive the necessary briefing on policies that relate to their management.

Refugee Status is not considered permanent when granted in these countries of asylum. Where the status of a refugee has to be reviewed every six months, it leaves the refugee in constant stress, thus perpetuating the already disturbed frame of mind.

The use of firearms to compel refugees in organized rural settlements to comply with rules and regulations is inhuman, especially under circumstances where refugees have already encountered crises in their lives. Moreover, this procedure changes the character of the refugee settlements to that of concentration camps.

Although this is an effort to enforce discipline and ensure the security of refugees, awareness of such policies lowers the morale of refugees to such an extent that they are afraid even to verbalize their personal problems. It also strains relationships and prevents the establishment of the necessary rapport between refugees and their supervisors. It also discourages them from responding enthusiastically to settlement activities. Everything planned for their assistance is suspected to have an ill-meaning motive.

Discipline can be effectively enforced in a non-military manner through tact, diplomacy, and incentives, and above all through education.

It is commendable that none of the policies under review pursue immediate deportation of refugees where this is recommended by the courts. Where persons awaiting deportation are given time to find a resettlement opportunity, it is believed that some agency would find travel documents and fund the journey out of the country. However, deportation as punishment of an asylum-seeker for having illegally entered the country makes refuge-seeking irrational. Where a person is desperate to save his life from imminent danger, his only concern is to reach a place of safety; whether that place is authorized or not is secondary. Refugees who are refused a status solely on account of illegal entry into a country of asylum nurse a feeling of bitterness against the government.

Restrictions in the movements of refugees along authorized routes are for their own protection. Nevertheless, it works against Convention requirements for freedom of movement and association with nationals.

'Settlement lock-ups' though understandably a local punitive measure, should be applied when offences are repeated in spite of counselling, advice or discussions, and definitely never applied to persons frail in body or mind or to women and children.

The nature of treatment refugees receive has been substantiated – whether justified or unjustified. Certain aspects of policies which promote the well-being of refugees are also worth considering.

As obtains in Botswana, there is merit in having a Refugee Advisory Committee responsible to the minister for the management of refugee affairs. In this respect the minister's discretionary powers are limited and his decisions are based on inputs made by experts serving on the Committee. The place of a Refugee Commissioner in all three countries also has merit as he serves as liaison between refugees and government. Exploiting these provisions in legislative instruments of all countries of asylum in Africa serves as an invaluable aspect of refugee administration.

Although membership of an Advisory Committee is not indicated in the policy, it is worthwhile making this a relevant aspect of legislation to include representatives of a fair cross-section of stakeholders in refugee affairs in the country – the security, the judiciary, the voluntary organizations, the operational partners and UNHCR, as well as other experts in health, education, agriculture and personnel management.

The deficiencies in the policies under review are best considered by the Advisory Committee under the following functions:

1 make rules and regulations for the guidance of refugee administrators and security personnel, including those at the borders, These rules will seek approval of the Minister for Refugee Affairs before they come into effect;

2 adopt measures for implementation of policies with particular reference to training of personnel, financing and cooperation with operational partners;

3 determine the criteria procedure for granting Refugee Status for guidance of committees and individuals assigned to screening of applicants for Refugee Status;

4 recognize asylum-seekers for Refugee Status and maintain a register of refugees;

5 specify the mode of application of each of the rights of refugees as enshrined in the UN Refugee Convention (1951);

6 issue residence permits, identity cards, work permits and other relevant documents for use by refugees;

7 determine measures of protection of all refugees in the various locations and oversee their implementation;

8 establish refugee settlements and make rules and regulations to ensure their maintenance and the protection of residents;

9 give guidance to operational partners in administration of refugee rights in areas, particularly health, education, and employment;

10 make the basic rules and regulations governing the award of scholarships to refugees by all voluntary organizations;

11 set up counselling services for the main purpose of education and rehabilitation of refugees;

12 participate in national programmes for disaster management, including emergency aid for large influxes of refugees;

13 supervise the welfare of refugees in police custody and review their cases at specific periods;

14 cooperate with UNHCR as the UN and OAU Refugee Conventions demand.

A special duty of an Advisory Committee is to monitor information that comes to its knowledge through its contact with refugees, that could form the basis for instability of the country and the uprootment of nationals and report this to the Minister of Refugee Affairs.

Preventing Refugee Crises

The United Nations High Commissioner for Refugees has repeatedly stated that 'The ultimate objective of international protection is to help a refugee to cease being a refugee'.

Preventive initiatives are taken in countries which produce refugees or which may do so in the future. Attempts are also made to forestall refugee flows through institution-building and training in countries at risk of producing refugees.

In situations where economically-motivated migrants may seek to take advantage of refugee channels, mass information programmes have been launched to provide a clearer understanding of refugee status. These programmes aim to discourage people who may seek to use asylum channels for economic reasons, while keeping them open for those who flee persecution.

Where civil strife and refugee crises have already erupted, humanitarian action does not only provide relief but it also averts further internal displacement to reduce the need for people to take refuge abroad. Humanitarian assistance has helped stabilize population movements and has eventually created conditions favourable to the return of refugees.

African countries are strongly implored to give more serious thought to the adoption of laws that aim to overcome the causes of uprootment as a means of prevention in addition to the cure of the ills.

Ultimately, the solution to the problems of refugees in Africa is not solely dependent on international and regional conventions but also on the collective conscience of African countries. In their efforts to solve the problems of refugees, African nations should hold high the well known dictum of Fridjof Nansen, first UN High Commissioner for Refugees (UNHCR Supplement No. 41, July 1973): that 'love of man is practical policy'.

Bibliography

All Africa Conference of Churches (1971), *Working Together in Helping Africa's Refugees*, AACC, Nairobi.

Brooks, H.C. and El-Ayouty, Y. (1970), *Refugees South of the Sahara: An African Dilemma*, Negro Universities Press, Westport, Connecticut.

Dennison, D. (1965), Social Policy and Administration Revisited, George Allen and Unwin Ltd, London.

Eyden, J.L.M. (1969), *Social Policy in Action*, Routledge and Kegan Paul Ltd, London.

Goedhert, G.J. van Heuven (1955), 'Refugee Problems and their Solutions', lecture given at Nobel Institute, Oslo, December.

Golan, N. (1969), 'When is a Client in Crisis?', *Social Casework*, New York.

Grahl-Madsen, *The Status of Refugees in International Law*, Volume II, *Asylum, Entry and Sojourn*.

Greenfield, R. (1982), *The OAU and Africa's Refugees*, Oxford University Press.

Gunther, J. (1955), *Inside Africa*, Hamish Hamilton Ltd, London.

Hamrell, S. (1967), *Refugee Problems in Africa*, The Scandinavian Institute of African Studies, Uppsala.

Hatch, J. (1960), *Africa Today and Tomorrow: an Outline of Basic Facts and Major Problems*, Frederick A. Praeger Inc., New York.

Hatch, J. (1965), *A History of Postwar Africa*, Frederick Praeger Inc., New York.

Holborn, L.W. (1975), *Refugees: A Problem of our Time (The Work of UNHCR 1951–72)*, M.J. Metucheu.

International Catholic Migration Commission (ICMC), *Migration News*, March/April 1970; July/August 1971, ICMC, Geneva.

International University Exchange Fund (IUEF), *Report on Problems of Refugees and Exiles in Europe*, IUEF, Geneva.

Khan, S.A. (1981), *Study on Human Rights and Massive Exodus*, UN, Geneva.

Livingstone, A. (1969), *Social Policy in Developing Countries*, Routledge and Kegan Paul Ltd, London.

Mair, L., *New Nations*, Weidenfeld and Nelson.

Mair, L., *Anthropology and Social Change*, Weidenfeld and Nelson.

Marshall, T.H. (1965), *Social Policy in the Twentieth Century*, Hutchinson and Co. Ltd., London.

Mazrui, A. (1969), *Violence and Thought*, Longmans, Green and Co. Ltd.

Melander, G. and Nobel, A. (1979), *International Legal Instruments on Refugees in Africa*, Scandinavian Institute of African Studies, Uppsala.

Miller, S.M. and Blessman, F. (1968), *Social Class and Social Policy*, Basic Books Inc., New York.

Nyerere, J.K. (1968), *UJAMAA (Essays on Socialism)*, Oxford University Press, London.

Nyerere, J.K. (1974), *Man and Development*, Oxford University Press, London.

Organization of African Unity (OAU) (1969), Convention Governing the Specific Aspects of Refugee Problems in Africa, OAU, Addis Ababa.

Pinker, R. (1971), *Social Theory and Social Policy*, Heinemann Educational Books Ltd. London.

Rapoport, L. (1962), 'The State of Crisis, Some Theoretical Considerations', *Social Services Review*, New York.

Rodney, W. (1972), *How Europe Underdeveloped Africa*, Tanzania Publishing House, Dar es Salaam.

Rooney, D.D. and Halliday E. (1966), *The Building of Modern Africa*, George Harcup and Co. Ltd, London.

Schoor, A.L. (1968), *Explorations in Social Policy*, Basic Books Inc., New York.

Sender, J. and Smith, S. (1986), *The Development of Capitalism in Africa*, Methuen, London.

Titmus, R.M. (1969), *Social Policy in Action*, Routledge and Kegan Paul Ltd, London.

UNECA (1976), *Survey of Economic and Social Conditions in Africa*, UN.

UNHCR (1955–89), *Refugees*, Public Information Section, UNHCR, Geneva.

UNHCR (1971), *A Mandate to Protect and Assist Refugees*, UN.

UNHCR, reports of UNHCR activities from 1973–89.

UNHCR, notes on International Protection 1974–1989, submitted to the annual meetings of the Executive Committee of the UNHCR programme.

UNHCR (1979a), *Collection of International Instruments concerning Refugees*, UNHCR, Geneva.

UNHCR (1979b), Report of the Conference on the Situations of Refugees in Africa, 7–17 May, Arusha, Tanzania, HCR/140/12/79, UNHCR, Geneva.

UNHCR, *Partnerships: A Programme Management Handbook for UNHCR*.

UNHCR, UNECA, OAU, Dag Hammerskjold Foundation (1968), Final Report of the Conference on the Legal, Economic and Social Aspects of African Refugee Problems, 9–18 October 1967.

United Nations (1970), *The World Social Situation*, UN, New York.

United Nations (1971), *As They Came in Africa*, UN, New York.

United Nations (1968), Convention and Protocol Relating to the Status of Refugees, UN.

Vernant, J.M. (1953), *The Refugee in the Post-War World*, T. and A. Constable, Edinburgh.

Wallerstein, I. (1971), Africa: The Politics of Independence, Vintage Books, New York.

Weis, P. (1954), *Legal Aspects of the Convention of 25th July, 1951 relating to the Status of Refugees*, Oxford University Press, London.

World Peace Through Law Centre (1976), report of the Committee on the International Legal Protection of Refugees, Washington DC.

Yeld, R. (1965), 'Implications of Experience with Refugee Settlement', East African Institute of Social Research Conference paper, Makerere University College.

Zwingmann, C.A. and Pfsiter, A.M. (1973), *Uprooting and After*, Springer-Verlag, New York.

APPENDICES

Appendix 1: Statute of the Office of the United Nations High Commissioner for Refugees

INTRODUCTION

In Resolution 319 (IV) of 3rd December, 1949, the United Nations General Assembly decided to establish a High Commissioner's Office for Refugees as of 1st January 1951.

The Statute of the Office of the United Nations High Commissioner for Refugees was adopted by the General Assembly on 14th December 1950 as Annex to Resolution 428(V). In this Resolution, reproduced on page 4, the Assembly also called upon Governments to co-operate with the High Commissioner in the performance of his functions concerning refugees falling under the competence of his Office. In accordance with the Statute, the work of the High Commissioner is humanitarian and social and of an entirely non-political character.

The functions of the High Commissioner are defined in the Statute and in various Resolutions subsequently adopted by the General Assembly. Resolutions concerning the High Commissioner's Office adopted by the General Assembly and the Economic and Social Council are issued by UNHCR as an information document, HCR/INF/48.

The High Commissioner reports annually to the General Assembly through the Economic and Social Council. Pursuant to paragraph 4 of the Statute, an Advisory Committee on Refugees was established by the Economic and Social Council[1] and was later reconstituted as the United Nations Refugee Fund (UNREF) Executive Committee.[2] The latter was replaced in 1958 by the Executive Committee of the High Commissioner's Programme.[3] Under its terms of reference, the Executive Committee, inter alia, approves and supervises material assistance programme of the High Commissioner's Office and advises the High Commissioner at his request on the exercise of his functions under the Statute. The Executive Committee was originally composed of 24 States. In 1963,[4] its membership was increased to 30 States so as to achieve the widest possible geographical representation.

The Office was originally established for a period of three years (Statute, paragraph 5). By General Assembly Resolutions 727 (VIII) of 23rd October 1953, 1165 (XII) of 26th November, 1957 and 1783 (XVII) of 7th December 1962, the Office was extended for successive periods of five years, the present term being until 31st December 1968.

The first United Nations High Commissioner for Refugees was the late Dr. G.J. van Heuven Goedhart, of the Netherlands (1951–1956) who was succeeded by Dr. A.R. Lindt (1957–1960 and Dr. Felix Schnyder (1961–1965), both of Switzerland. The present High Commissioner is Prince Sadruddin Aga Khan (Iran) who assumed office on 1st January, 1966.

The Headquarters of the High Commissioner's Office are located at Geneva, Switzerland. The High Commissioner has appointed representatives and correspondents in a number of areas throughout the five continents.

Geneva, December 1966.

GENERAL ASSEMBLY RESOLUTION 428 (V)
OF 14TH DECEMBER 1950

The General Assembly,

In view of its resolution 319 a) (IV) of 3rd December, 1949,

1. Adopts the annex to the present resolution, being the Statute of the Office of the United Nations High Commissioner for Refugees;

2. Calls upon Governments to co-operate with the United Nations High Commissioner for Refugees in the performance of his functions concerning refugees falling under the competence of his Office, especially by:

   a) becoming parties to international conventions providing for the protection of refugees, and taking the necessary steps of implementation under such conventions;

   b) Entering into special agreements with the High Commissioner for the executive of measures calculated to improve the situation of refugees and to reduce the number requiring protection;

   c) Admitting refugees to their territories, not excluding those in the most destitute categories;

   d) Assisting the High Commissioner in his efforts to promote the voluntary repatriation of refugees;

e) Promoting the assimilation of refugees, especially be facilitating their naturalization;

f) Providing refugees with travel and other documents such as would normally be provided to other aliens by their national authorities, especially documents which would facilitate their resettlement;

g) Permitting refugees to transfer their assets and especially those necessary for their resettlement;

h) Providing the High Commissioner with information concerning the number of condition of refugees and laws and regulations concerning them.

3. Requests the Secretary-General to transmit the present resolution, together with the annex attached thereto, also to States non-members of the United Nations, with a view to obtaining their co-operation in its implementation.

ANNEX
CHAPTER I: GENERAL PROVISIONS

1.   The United Nations High Commissioner for Refugees, acting under the authority of the General Assembly, shall assume the function of providing international protection, under the auspices of the United Nations, to refugees who fall within the scope of the present Statute and of seeking permanent solutions for the problem of refugees by assisting Governments and, subject to the approval of the Governments concerned, private organizations to facilitate the voluntary repatriation of such refugees, or their assimilation within new national communities.

In the exercise of his functions, more particularly when difficulties arise, and for instance with regard to any controversy concerning the international status of these persons, the High Commission shall request the opinion of the advisory committee on refugees if it is created.

2.   The work of the High Commissioner shall be of an entirely non-political character; it shall be humanitarian and social and shall relate, as a rule, to groups and categories of refugees.

3.   The High Commissioner shall follow policy directives given him by the General Assembly or the Economic and Social Council.

4.   The Economic and Social Council may decide, after hearing the views of the High Commissioner on the subject, to establish an advisory committee on refugees, which shall consist of representatives of States Members and States non-members of the United Nations, to be selected by the Council on

the basis of their demonstrated interest in and devotion to the solution of the refugee problem.

5.   The General Assembly shall review, not later than at its eighth regular session, the arrangements for the Office of the High Commissioner with a view to determining whether the Office should be continued beyond 31st December 1953.

CHAPTER II: FUNCTIONS OF THE HIGH COMMISSIONER

6.   The competence[5] of the High Commissioner shall extent to:

A.   i) Any person who has been considered a refugee under the Arrangements of 12th May 1926 and of 30th June 1928 or under the Conventions of 28th October 1933 and 10th February 1938, the Protocol of 14th September 1939 or the Constitution of the International Refugee Organization.

ii) Any person who, as a result of events occurring before 1st January 1951 and owing to well-founded fear of being persecuted for reasons or race, religion, nationality or political opinion, is outside the country of his nationality and is unable or, owing to such fear or for reasons other than personal convenience, is unwilling to avail himself of the protection of that country; or who, not having a nationality and being outside the country of his former habitual residence, is unable or, owing to such fear or for reasons other than personal convenience, is unwilling to return to it. Decisions as to eligibility taken by the International Refugee Organization during the period of its activities shall not prevent the status of refugee being accorded to persons who fulfill the conditions of the present paragraph;

The competence of the High Commissioner shall cease to apply to any person defined in section (a) above if;

a) he has voluntarily reavailed himself of the protection of the country of his nationality; or

b) having lost his nationality, he has voluntarily reacquired it; or

c) he has acquired a new nationality, and enjoys the protection of the country of his new nationality; or

d) he has voluntarily re-established himself in the country which he left or outside which he remained owing to fear of persecution; or

e) he can no longer, because the circumstances in connection with which he has been recognized as a refugee have ceased to exist, claim grounds other than those of personal convenience for continuing to refuse to avail himself of the protection of the country of his nationality. Reasons of a purely economic character may not be invoked; or

f) being a person who has no nationality, he can no longer, because the circumstances in connection with which he has been recognized as a refugee have ceased to exist and he is able to return to the country of his former habitual residence, claim grounds other than those of personal convenience for continuing to refuse to return to that country;

B.   Any other person who is outside the country of his nationality, or if he has no nationality, the country of his former habitual residence, because he has or had well-founded fear of persecution by reason of his race, religion, nationality or political opinion and is unable or, because of such fear, is unwilling to avail himself of the protection of the government of the country of his nationality, or, if he has no nationality, to return to the country of his former habitual residence.

7.   Provided that the competence of the High Commissioner as defined in paragraph 6 above shall not extend to a person;

a) who is a national of more than one country unless he satisfies the provisions of the preceding paragraph in relation to each of the countries of which he is a national; or

b) who is recognized by the competent authorities of the country in which he has taken residence as having the rights and obligations which are attached to the possession of the nationality of that country; or

c) who continues to receive from other organs or agencies of the United Nations protection or assistance; or

d) in respect of whom there are serious reasons for considering that he has committed a crime covered by the provisions of treaties of extradition or a crime mentioned in Article VI of the London Charter of the International Military Tribunal or by the provisions of Article 14, paragraph 2, of the Universal Declaration of Human Rights.[6]

8.   The High Commissioner shall provide for the protection of refugees falling under the competence of his Office by:

a) promoting the conclusion and ratification of international conventions for the protection of refugees, supervising their application and proposing amendments thereto;

b) promoting through special agreements with Governments the execution of any measures calculated to improve the situation of refugees and to reduce the number requiring protection;

c) assisting governmental and private efforts to promote voluntary repatriation or assimilation within new national communities;

d) promoting the admission of refugees, not excluding those in the most destitute categories, to the territories of States;

e) endeavouring to obtain permission for refugees to transfer their assets and especially those necessary for their resettlement;

f) obtaining from Governments information concerning the number and conditions of refugees in their territories and the laws and regulations concerning them;

g) keeping in close touch with the Governments and inter-governmental organizations concerned;

h) establishing contact in such manner as he may think best with private organizations dealing with refugee questions;

i) facilitating the co-ordination of the efforts of private organizations concerned with the welfare of refugees.

9.   The High Commissioner shall engage in such additional activities, including repatriation and resettlement, as the General Assembly may determine, within the limits of the resources placed at his disposal.

10.   The High Commissioner shall administer any funds, public or private, which he receives for assistance to refugees, and shall distribute them among the private and, as appropriate, public agencies which he deems best qualified to administer such assistance.

The High Commissioner may reject any offers which he does not consider appropriate or which cannot be utilized.

The High Commissioner shall not appeal to Governments for funds or make a general appeal, without the prior approval of the General Assembly.

The High Commissioner shall include in his annual report a statement of his activities in this field.

11.   The High Commissioner shall be entitled to present his views before the General Assembly, the Economic and Social Council and their subsidiary bodies.

The High Commissioner shall report annually to the General Assembly through the Economic and Social Council; his report shall be considered as a separate item on the agenda of the General Assembly.

12.   The High Commissioner may invite the co-operation of the various specialized agencies.

CHAPTER III: ORGANIZATION AND FINANCES

13.   The High Commissioner shall be elected by the General Assembly on the nomination of the Secretary-General. The terms of appointment of the High Commissioner shall be proposed by the Secretary-General and approved by the General Assembly. The High Commissioner shall be elected for a term of three years, from 1st January 1951.

14.   The High Commissioner shall appoint, for the same term a Deputy High Commissioner of a nationality other than his own.

15.   a)   Within the limits of the budgetary appropriations provided, the staff of the Office of the High Commissioner shall be appointed by the High Commissioner and shall be responsible to him in the exercise of their functions.

     b)   Such staff shall be chosen from persons devoted to the purposes of the Office of the High Commissioner.

     c)   Their conditions of employment shall be those provided under the staff regulations adopted by the General Assembly and the rules promulgated thereunder by the Secretary-General.

     d)   Provision may also be made to permit the employment of personnel without compensation.

16.   The High Commissioner shall consult the Government of the countries of residence of refugees as to the need for appointing representatives therein. In any country recognizing such need, there may be appointed a representative approved by the Government of that country. Subject to the foregoing, the same representative may serve in more than one country.

17.   The High Commissioner and the Secretary-General shall make appropriate arrangements for liaison and consultation on matters of mutual interest.

18.   The Secretary-General shall provide the High Commissioner with all necessary facilities within budgetary limitations.

19.   The Office of the High Commissioner shall be located in Geneva, Switzerland.

20.   The Office of the High Commissioner shall be financed under the budget of the United Nations. Unless the General Assembly subsequently decided otherwise, no expenditure other than administrative expenditures relating to the functioning of the Office of the High Commissioner shall be borne on the budget of the United Nations and all other expenditures relating to the activities of the High Commissioner shall be financed by voluntary contributions.

21.   The administration of the Office of the High Commissioner shall be subject to the Financial Regulations of the United Nations and to the financial rules promulgated thereunder by the Secretary-General.

22.   Transactions relating to the High Commissioner's funds shall be subject to audit by the United Nations Board of Auditors, provided that the Board may accept accounts from the agencies to which funds have been allocated. Administrative arrangements for the custody of such funds and their allocation shall be agreed between the High Commissioner and the Secretary-General in accordance with the Financial Regulations of the United Nations and rules promulgated thereunder by the Secretary-General.

Notes

1   Resolution 393 (XIII) B of 10th September, 1951.
2   Economic and Social Council Resolution 565 (XIX) of 31st March 1955 adopted pursuant to General Assembly Resolution 832 (IX) of 21st October, 1954.
3   General Assembly Resolution 1166 (XII) of 26th November, 1957 and Economic and Social Council Resolution 672 (XXV) of 30th April, 1958.
4   General Assembly Resolution 1958 (XVIII) of 12th December, 1963.
5   In addition to refugees as defined in the Statute, other categories of persons finding themselves in refugee-like situations, have in the course of the years come within the concern of the High Commissioner in accordance with the subsequent General Assembly and ECOSOC Resolutions.
6   See UN General Assembly Resolution 217 A (III) of 10th December 1948.

Appendix 2: Convention Relating to the Status of Refugees, 1951[1]

PREAMBLE

The High Contracting Parties

Considering that the Charter of the United Nations and the Universal Declaration of Human Rights approved on 10th December 1948 by the General Assembly have affirmed the principle that human beings shall enjoy fundamental rights and freedoms without discrimination,

Considering that the United Nations has, on various occasions, manifested its profound concern for refugees and endeavoured to assure refugees the widest possible exercise of these fundamental rights and freedoms,

Considering that it is desirable to revise and consolidate previous international agreements relating to the status of refugees and to extend the scope of and the protection accorded by such instruments by means of a new agreement,

Considering that the grant of asylum may place unduly heavy burdens on certain countries, and that a satisfactory solution of a problem of which the United Nations has recognized the international scope and nature cannot therefore be achieved without international co-operation,

Expressing the wish that all States, recognizing the social and humanitarian nature of the problem of refugees, will do everything within their power to prevent this problem from becoming a cause of tension between States,

Noting that the United Nations High Commissioner for Refugees is charged with the task of supervising international conventions providing for the protection of refugees, and recognizing that the effective co-ordination of measures taken to deal with this problem will depend upon the co-operation of States with the High Commissioner,

Having agreed as follows:

CHAPTER I: GENERAL PROVISIONS

Article 1
DEFINITION OF THE TERM 'REFUGEE'

A.   For the purposes of the present Convention, the term 'refugee' shall apply to any person who:

1) has been considered a refugee under the Arrangements of 12th May 1926 and 30th June 1928 or under the Conventions of 28th October 1933 and 10th February 1938, the Protocol of 14th September 1939 or the Constitution of the International Refugee Organization;

Decisions of non-eligibility taken by the International Refugee Organization during the period of its activities shall not prevent the status of refugee being accorded to persons who fulfill the conditions of paragraph 2 of this section;

2) As a result of events occurring before 1st January 1951 and owing to well-founded fear of being persecuted for reasons of race, religion, nationality, membership of a particular social group or political opinion, is outside the country of his nationality and is unable or, owing to such fear, is unwilling to avail himself of the protection of that country; or who, not having a nationality and being outside the country of his former habitual residence as a result of such events, is unable or, owing to such fear, is unwilling to return to it.

In the case of a person who has more than one nationality, the term 'the country of his nationality' shall mean each of the countries of which he is a national, and a person shall not be deemed to be lacking the protection of the country of his nationality if, without any valid reason based on well-founded fear, he has not availed himself of the protection of one of the countries of which he is a national.

B. 1) For the purposes of this Convention, the words 'events occurring before 1st January 1951' in Article 1, Section A, shall be understood to mean either

a) 'events occurring in Europe before 1st January 1951'; or

b) 'events occurring in Europe or elsewhere before 1st January 1951'; and each Contracting State shall make a declaration at the time of signature, ratification or accession, specifying which of these meanings it applies for the purpose of its obligations under this Convention.

2) Any Contracting State which has adopted alternative (a) may at any time extend its obligations by adopting alternative (b) by means of a notification addressed to the Secretary-General of the United Nations.

C.   This Convention shall cease to apply to any person falling under the terms of section A. if;

1) He has voluntarily re-availed himself of the protection of the country of his nationality; or

2) Having lost his nationality, he has voluntarily re-acquired it; or

3) He has acquired a new nationality, and enjoys the protection of the country of his new nationality; or

4) He has voluntarily re-established himself in the country which he left or outside which he remained owing to fear of persecution; or

5) He can no longer, because the circumstances in connexion with which he has been recognized as a refugee have ceased to exist, continue to refuse to avail himself of the protection of the country of his nationality;

Provided that this paragraph shall not apply to a refugee falling under section A 1) of this Article who is able to invoke compelling reasons arising out of previous persecution for refusing to avail himself of the protection of the country of nationality;

6) Being a person who has no nationality he is, because the circumstances in connexion with which he has been recognized as a refugee have ceased to exist, able to return to the country of his former habitual residence;

Provided that this paragraph shall not apply to a refugee falling under section A (1) of this Article who is able to invoke compelling reasons arising out of previous persecution for refusing to return to the country of his former habitual residence.

D.   The Convention shall not apply to persons who are at present receiving from organs or agencies of the United Nations other than the United Nations High Commissioner for Refugees protection or assistance.

When such protection of assistance has ceased for any reason, without the position of such persons being definitely settled in accordance with the relevant resolutions adopted by the General Assembly of the United Nations, these persons shall ipso facto be entitled to the benefits of this Convention.

E.   This Convention shall not apply to a person who is recognized by the competent authorities of the country in which he has taken residence as having the rights and obligations which are attached to the possession of the nationality of that country.

F.   The provisions of this Convention shall not apply to any person with respect to whom there are serious reasons for considering that;

(a) he has committed a crime against peace, a war crime, or a crime against humanity, as defined in the international instruments drawn up to make provision in respect of such crimes;

(b) he has committed a serious non-political crime outside the country of refuge prior to his admission to that country as a refugee;

(c) he has been guilty of acts contrary to the purposes and principles of the United Nations.

Article 2
GENERAL OBLIGATIONS

Every refugee has duties to the country in which he finds himself, which require in particular that he conform to its laws and regulations as well as to measures taken for the maintenance of public order.

Article 3
NON-DISCRIMINATION

The Contracting States shall apply the provisions of this Convention to refugees without discrimination as to race, religion or country of origin.

Article 4
RELIGION

The Contracting States shall accord to refugees within their territories treatment at least as favourable as that accorded to their nationals with respect to freedom to practise their religion and freedom as regards the religious education of their children.

Article 5
RIGHTS GRANTED APART FROM THIS CONVENTION

Nothing in this Convention shall be deemed to impair any rights and benefits granted by the Contracting State to refugees apart from this Convention.

Article 6
THE TERM 'IN THE SAME CIRCUMSTANCES'

For the purposes of this Convention, the term 'in the same circumstances' implies that any requirements (including requirements as to length and conditions of sojourn or residence) which the particular individual would have to fulfil for the enjoyment of the right in question, if he were not a refugee, must be fulfilled by him, with the exception of requirements which by their nature a refugee is incapable of fulfilling.

Article 7
EXEMPTION FROM RECIPROCITY

1.   Expect where this convention contains more favourable provisions, a Contracting State shall accord to refugees the same treatment as is accorded to aliens generally.

2.   After a period of three years' residence, all refugees shall enjoy exemption from legislative reciprocity in the territory of the Contracting States.

3.   Each Contracting State shall continue to accord to refugees the rights and benefits to which they were already entitled, in the absence of reciprocity, at the date of entry into force of this Convention for that State.

4.   The Contracting States shall consider favourably the possibility of according to refugees, in the absence of reciprocity, rights and benefits beyond those to which they are entitled according to paragraphs 2 and 3, and to extending exemption from reciprocity to refugees who do not fulfil the conditions provided for in paragraphs 2 and 3.

5.   The provisions of paragraphs 2 and 3 apply both to the rights and benefits referred to in Articles 13, 18, 19, 21 and 22 of this Convention and to rights and benefits for which this Convention does not provide.

Article 8
EXEMPTION FROM EXCEPTIONAL MEASURES

With regard to exceptional measures which may be taken against the person, property or interests of nationals of a foreign State, the Contracting States shall not apply such measures to a refugee who is formally a national of the said State solely on account of such nationality. Contracting States, which, under their legislation, are prevented from applying the general principle expressed in this article, shall, in appropriate cases, grant exemptions in favour of such refugees.

Article 9
PROVISIONAL MEASURES

Nothing in this Convention shall prevent a Contracting State, in time of war or other grave and exceptional circumstances, from taking provisionally measures which it considers to be essential to the national security in the case of a particular person, pending a determination by the Contracting State that that person is in fact a refugee and that the continuance of such measures is necessary in his case in the interests of national security.

Article 10
CONTINUITY OF RESIDENCE

1.   Where a refugee has been forcibly displaced during the Second World War and removed to the territory of a Contracting State, and is resident there, the period of such enforced sojourn shall be considered to have been lawful residence within that territory.

2.   Where a refugee has been forcibly displaced during the Second World War from the territory of a Contracting State and has, prior to the date of entry into force of this Convention, returned there for the purpose of taking up residence, the period of residence before and after such enforced displacement shall be regarded as one uninterrupted period for any purposes for which uninterrupted residence is required.

Article 11
REFUGEE SEAMEN

In the case of refugees regularly serving as crew members on board a ship flying the flag of a Contracting State, that state shall given sympathetic consideration to their establishment on its territory and the issue of travel documents to them or their temporary admission to its territory and the issue of travel documents to them or their temporary admission to its territory particularly with a view to facilitating their establishment in another country.

CHAPTER II: JURIDICAL STATUS

Article 12
PERSONAL STATUS

1.   The personal status of a refugee shall be governed by the law of the country of his domicile or, if he has no domicile, by the law of the country of his residence.

2.   Rights previously acquired by a refugee and dependent on personal status, more particularly rights attaching to marriage, shall be respected by a Contracting State, subject to compliance, if this be necessary, with the formalities required by the law of that State, provided that the right in question is one which would have been recognized by the law of that State had be not become a refugee.

Article 13
MOVABLE AND IMMOVABLE PROPERTY

The Contracting States shall accord to a refugee treatment as favourable as possible and, in any event, not less favourable than that accorded to aliens generally in the same circumstances, as regards the acquisition of movable and immovable property and other rights pertaining thereto, and to leases and other contracts relating to movable and immovable property.

Article 14
ARTISTIC RIGHTS AND INDUSTRIAL PROPERTY

In respect of the protection of industrial property, such as inventions, designs or models, trade marks, trade names, and of rights in literacy, artistic and scientific works, a refugee shall be accorded in the country in which he has his habitual residence the same protection as is accorded to nationals of that country. In the territory of any other Contracting State, he shall be accorded the same protection as is accorded in that territory to nationals of the country in which he has his habitual residence.

Article 15
RIGHT OF ASSOCIATION

As regards non-political and non-profit-making associations and trade unions the Contracting States shall accord to refugees lawfully staying in their territory the most favourable treatment accorded to nationals of a foreign country, in the same circumstances.

Article 16
ACCESS TO COURTS

1.   A refugee shall have free access to the courts of law in the territory of all Contracting States.

2.   A refugee shall enjoy in the Contracting State in which he has his habitual residence the same treatment as a national in matters pertaining to access to the Courts, including legal assistance and exemption from cautio judicatum solvi.

3.   A refugee shall be accorded in the matters referred to in paragraph 2 in countries other than that in which he has his habitual residence the treatment granted to a national of the country of his habitual residence.

CHAPTER III: GAINFUL EMPLOYMENT

Article 17
WAGE-EARNING EMPLOYMENT

1.   The Contracting States shall accord to refugees lawfully staying in their territory the most favourable treatment accorded to nationals of a foreign country in the same circumstances, as regards the right to engage in wage-earning employment.

2.   In any case, restrictive measures imposed on aliens or the employment of aliens for the protection of the national labour market shall not be applied to a refugee who was already exempt from them at the date of entry into force of this Convention for the Contracting State concerned, or who fulfils one of the following conditions:

(a) He has completed three years' residence in the country;

(b) He has a spouse possessing the nationality of the country of residence. A refugee may not invoke the benefits of this provision if he has abandoned his spouse.

(c) He has one or more children possessing the nationality of the country of residence.

3.   The Contracting States shall give sympathetic consideration to assimilating the rights of all refugees with regard to wage-earning employment to those of nationals, and in particular of those refugees who have entered their territory pursuant to programmes of labour recruitment or under immigration schemes.

Article 18
SELF-EMPLOYMENT

The Contracting States shall accord to a refugee lawfully in their territory treatment as favourable as possible and, in any event, not less favourable than that accorded to aliens generally in the same circumstances, as regards the right to engage on his own account in agriculture, industry, handicrafts and commerce and to establish commercial and industrial companies.

Article 19
LIBERAL PROFESSIONS

1.   Each Contracting State shall accord to refugees lawfully staying in their territory who hold diplomas recognized by the competent authorities of that State, and who are desirous of practicing a liberal profession, treatment as favourable as possible and, in any event, not less favourable than that accorded to aliens generally in the same circumstances.

2.   The Contracting States shall use their best endeavours consistently with their laws and constitutions to secure the settlement of such refugees in the territories, other than the metropolitan territory, for whose international relations they are responsible.

CHAPTER IV: WELFARE

Article 20
RATIONING

Where a rationing system exists, which applies to the population at large and regulates the general distribution of products in short supply, refugees shall be accorded the same treatment as nationals.

Article 21
HOUSING

As regards housing, the Contracting States, in so far as the matter is regulated by laws or regulations or is subject to the control of public authorities, shall accord to refugees lawfully staying in their territory treatment as favourable as possible and, in any event, not less favourable than that accorded to aliens generally in the same circumstances.

Article 22
PUBLIC EDUCATION

1.   The Contracting States shall accord to refugees the same treatment as us accorded to nationals with respect to elementary education.
2.   The Contracting States shall accord to refugees treatment as favourable as possible, and, in any event, not less favourable than that accorded to aliens generally in the same circumstances, with respect to education other than elementary education and, in particular, as regards access to studies, the recognition of foreign school certificates, diplomas and degrees, the remission of fees and charges and the award of scholarships.

Article 23
PUBLIC RELIEF

The Contracting States shall accord to refugees lawfully staying in their territory the same treatment with respect to public relief and assistance as is accorded to their nationals.

Article 24
LABOUR LEGISLATION AND SOCIAL SECURITY

1.   The Contracting States shall accord to refugees lawfully staying in their country the same treatment as is accorded to nationals in respect of the following matters:
(a) In so far as such matters are governed by laws or regulations or are subject to the control of administrative authorities: remuneration, including family allowances where these form part of remuneration, hours of work, overtime arrangements, holidays with pay, restrictions on home work, minimum age of employment, apprenticeship and training, women's work and the

work of young persons, and the enjoyment of the benefits of collective bargaining;

(b) Social security (legal provisions in respect of employment injury, occupational diseases, maternity, sickness, disability, old age, death, unemployment, family responsibilities and any other contingency which, according to national laws or regulations, is covered by a social security scheme), subject to the following limitations:

(i) there may be appropriate arrangements for the maintenance of acquired rights and rights in course of acquisition;

(ii) National laws or regulations of the country of residence may prescribe special arrangements concerning benefits or portions or benefits which are payable wholly out of public funds, and concerning allowances paid to persons who do not fulfil the contributions conditions prescribed for the award of a normal pension.

2.   The right to compensation for the death of a refugee resulting from employment injury of from occupational disease shall not be affected by the fact that the residence of the beneficiary is outside the territory of the Contracting State.

3.   The Contracting States shall extend to refugees the benefits of agreements concluded between them, or which may be concluded between them in the future, concerning the maintenance of acquired rights and rights in the process of acquisition in regard to social security, subject only to the conditions which apply to nationals of the states signatory to the agreements in question.

4.   The Contracting States will give sympathetic consideration to extending to refugees so far as possible the benefits of similar agreements which may at any time be in force between such Contracting States and non-Contracting States.

CHAPTER V: ADMINISTRATIVE MEASURES

Article 25
ADMINISTRATIVE ASSISTANCE

1.   When the exercise of a right by a refuges would normally require the assistance of authorities of a foreign country to whom he cannot have recourse, the Contracting States in whose territory he is residing shall arrange that such assistance be afforded to him by their own authorities or by an international authority.

2.   The authority or authorities mentioned in paragraph 1 shall deliver or cause to be delivered under their supervision to refugees such documents or certifications as would normally be delivered to aliens by or through their national authorities.

3.   Documents or certifications so delivered shall stand in the stead of the official instruments delivered to aliens by or through their national authorities, and shall be given credence in the absence of proof to the contrary.

4.   Subject to such exceptional treatment as may be granted to indigent persons, fees may be charged for the services mentioned herein, but such fees shall be moderate and commensurate with those charged to nationals for similar services.

5.   The provisions of this article shall be without prejudice to articles 27 and 28.

Article 26
FREEDOM OF MOVEMENT

Each Contracting State shall accord to refugees lawfully in its territory the right to choose their place of residence and to move freely within its territory, subject to any regulations applicable to aliens generally in the same circumstances.

Article 27
IDENTITY PAPERS

The Contracting States shall issue Identity Papers to any refugee in their territory who does not possess a valid travel document.

Article 28
TRAVEL DOCUMENTS

1.   The Contracting States shall issue to refugees lawfully staying in their territory travel documents for the purpose of travel outside their territory unless compelling reasons of national security or public order otherwise require, and the provisions of the Schedule to this Convention shall apply with respect to such documents. The Contracting States may issue such a travel document to any other refugee in their territory; they shall in particular give sympathetic consideration to the issue of such a travel document to refugees in their territory

who are unable to obtain a travel document from the country of their lawful residence.

2.   Travel documents issued to refugees under previous international agreements by parties thereto shall be recognized and treated by the Contracting States in the same way as if they had been issued pursuant to this article.

Article 29
FISCAL CHARGES

1.   The Contracting State shall not impose upon refugees duties, charges or taxes, or any description whatsoever, other or higher than those which are or may be levied on their nationals in similar situations.

2.   Nothing in the above paragraph shall prevent the application to refugees of the laws and regulations concerning charges in respect of the issue to aliens of administrative documents including identity papers.

Article 30
TRANSFER OF ASSETS

1.   A Contracting State shall, in conformity with its laws and regulations, permit refugees to transfer assets which they have brought into its territory, to another country where they have been admitted for the purposes of resettlement.

2.   A Contracting State shall give sympathetic consideration to the application of refugees for permission to transfer assets wherever they may be and which are necessary for their resettlement in another country to which they have been admitted.

Article 31
REFUGEES UNLAWFULLY IN THE COUNTRY OF REFUGE

1.   The Contracting States shall not impose penalties, on account of their illegal entry or presence, on refugees who, coming directly from a territory where their life or freedom was threatened in the sense of Article 1, enter or are present in their territory without authorization, provided they present themselves without delay to the authorities and show good cause for their illegal entry or presence.

2.   The Contracting States shall not apply to the movements of such refugees restrictions other than those which are necessary and such restrictions

shall only be applied until their status in the country is regularized or they obtain admission into another country. The Contracting States shall allow such refugees a reasonable period and all the necessary facilities to obtain admission into another country.

Article 32
EXPULSION

1.   The Contracting States shall not expel a refugee lawfully in their territory save on grounds of national security or public order.

2.   The expulsion of such a refugee shall be only in pursuance of a decision reached in accordance with due process of law. Except where compelling reasons of national security otherwise require, the refugee shall be allowed to submit evidence to clear himself, and to appeal to and be represented for the purpose before competent authority or a person or persons specially designated by the competent authority.

3.   Contracting States shall allow such a refugee a reasonable period within which to seek legal admission into another country. The Contracting States reserve the right to apply during that period such internal measures as they may deem necessary.

Article 33
PROHIBITION OF EXPULSION OR RETURN ('REFOULEMENT')

1.   No Contracting State shall expel or return ('refouler') a refugee in any manner whatsoever to the frontiers of territories where his life or freedom would be threatened on account of his race, religion, nationality, membership of a particular social group or political opinion.

2.   The benefit of the present provision may not, however be claimed by a refugee whom there are reasonable grounds for regarding as a danger to the security of the country in which he is, or who, having been convicted by a final judgement of a particularly serious crime, constitutes a danger to the community of that country.

Article 34
NATURALIZATION

The Contracting States shall as far as possible facilitate the assimilation and naturalization of refugees. They shall in particular make every effort to expedite naturalization proceedings and to reduce as far as possible the charges and costs of such proceedings.

CHAPTER VI: EXECUTORY AND TRANSITORY PROVISIONS

Article 35
CO-OPERATION OF THE NATIONAL AUTHORITIES WITH THE UNITED NATIONS

1.   The Contracting States undertake to co-operate with the office of the United Nations High Commissioner for Refugees, or any other agency of the United Nations which may succeed it, in the exercise of its functions, and shall in particular facilitate its duty of supervising the application of the provisions of this Convention.

2.   In order to enable the office of the High Commissioner or any other agency of the United Nations which may succeed it, to make reports to the competent organs of the United Nations, the Contracting States undertake to provide them in the appropriate form with information and statistical data requested concerning:

(a) the condition of refugees;

(b) the implementation of this Convention, and

(c) laws, regulations and decrees which are, or may hereafter be, in force relating to refugees.

Article 36
INFORMATION ON NATIONAL LEGISLATION

The Contracting States shall communicate to the Secretary-General of the United Nations the laws and regulations which they may adopt to ensure the application of this Convention.

Article 37
RELATION TO PREVIOUS CONVENTIONS

Without prejudice to article 28, paragraph 2, of this Convention, this Convention replaces, as between parties to it, the Arrangements of 5th July 1922, 31st May 1924, 12th May 1926, 30th June 1928 and 30th July 1935, the Conventions of 28th October 1933 and 19th February 1938, the Protocol of 14th September 1939 and the Agreement of 15th October 1946.

CHAPTER VII: FINAL CLAUSES

Article 38
SETTLEMENT OF DISPUTES

Any dispute between parties to this Convention relating to its interpretation or application, which cannot be settled by other means, shall be referred to the International Court of Justice at the request of any one of the parties to the dispute.

Article 39
SIGNATURE, RATIFICATION, AND ACCESSION

1.　　This Convention shall be opened for signature at Geneva on 28th July 1961 and shall thereafter be deposited with the Secretary-General of the United Nations. It shall be open for signature at the European Office of the United Nations from 28th July to 31st August 1951 and shall be reopened for signature at the Headquarters of the United Nations from 17th September 1951 to 31st December 1952.

2.　　This Convention shall be open for signature on behalf of all States Members of the United Nations, and also on behalf of any other state invited to attend to conference of plenipotentiaries on the Status of Refugees and Stateless Persons or to which an invitation to sign will have been addressed by the General Assembly. It shall be ratified and the instruments of ratification shall be deposited with the Secretary-General of the United Nations.

3.　　This Convention shall be open from 28th July 1951 for accession by the States referred to in paragraph 2 of this Article. Accessions shall be effected by the deposit of an instrument of accession with the Secretary-General of the United Nations.

Article 40
TERRITORIAL APPLICATION CLAUSE

1.　　Any State may, at the time of signature, ratification or accession, declare that this Convention shall extend to all or any of the territories for the international relations of which it is responsible. Such a declaration shall take effect when the Convention enters into force for the State concerned.

2.　　At any time thereafter any such extension shall be made by notification addressed to the Secretary-General of the United Nations and shall take effect as from the ninetieth day after the day of receipt by the Secretary-General of the United Nations of this notification, or as from the date of entry into force of the Convention for the State concerned, whichever is the later.

3.   With respect to those territories to which this Convention is not extended at the time of signature, ratification or accession, each state concerned shall consider the possibility of taking the necessary steps in order to extend the application of this Convention to such territories, subject, where necessary for constitutional reasons, to the consent of the governments of such territories.

Article 41
FEDERAL CLAUSE

In the case of a Federal or non-unitary State, the following provisions shall apply:

(a) with respect to those articles of this Convention that come within the legislative jurisdiction of the federal legislative authority, the obligations of the Federal Government shall to this extent be the same as those of parties which are not Federal States;

(b) With respect to those articles of this Convention that some within the legislative jurisdiction of constituent states, provinces or cantons which are not, under the constitutional system of the federation, bound to take legislative action, the Federal Government shall bring such articles with a favourable recommendation to the notice of the appropriate authorities of states, provinces or cantons at the earliest possible moment.

(c) A Federal State party to this Convention shall, at the request of any other Contracting State transmitted through the Secretary-General of the United Nations, supply a statement of the law and practice of the Federation and its constituent units in regard to any particular provision of the Convention showing the extent to which effect has been given to that provision by legislative or other action.

Article 42
RESERVATIONS

1.   At the time of signature, ratification or accession, any State may make reservations to articles of the Convention other than to articles 1, 3, 4, 16 (1), 33, 36–46 inclusive.

2.   Any State making a reservation in accordance with paragraph 1 of this article may at any time withdraw the reservation by a communication to that effect addressed to the Secretary-General of the United Nations.

Article 43
ENTRY INTO FORCE

1.   This Convention shall come into force on the ninetieth day following the day of deposit of the sixth instrument of ratification or accession.
2.   For each State ratifying or acceding to the Convention after the deposit of the sixth instrument of ratification or accession, the Convention shall enter into force on the ninetieth day following the date of deposit by such State of its instrument of ratification or accession.

Article 44
DENUNCIATION

1.   Any Contracting State may denounce this Convention at any time by a notification addressed to the Secretary-General of the United Nations.
2.   Such denunciation shall take effect for the Contracting State concerned one year from the date upon which it is received by the Secretary-General of the United Nations.
3.   Any State which has made a declaration or notification under article 40 may, at any time thereafter, by a notification to the Secretary-General of the United Nations, declare that the Convention shall cease to extend to such territory one year after the date of receipt of the notification by the Secretary-General.

Article 45
REVISION

1.   Any Contracting States may request revision of this Convention at any time by a notification addressed to the Secretary-General of the United Nations.
2.   The General Assembly of the United Nations shall recommend the steps, if any, to be taken in respect of such a request.

Article 46
NOTIFICATIONS BY THE SECRETARY-GENERAL
OF THE UNITED NATIONS

The Secretary-General of the United Nations shall inform all members of the United Nations and non-member States referred to in Article 39:
a) of declarations and notifications in accordance with Section B of article 1;
b) of signatures, ratifications and accessions in accordance with article 39;
c) of declarations and notifications in accordance with article 40;
d) of reservations and withdrawals in accordance with article 42;
e) of the date on which this Convention will come into force in accordance with article 43;
f) of denunciations and notifications in accordance with article 44;
g) of requests for revision in accordance with article 45.

In faith whereof the undersigned, duly authorized, have signed this Convention on behalf of their respective Governments,
Done at Geneva, this Twenty-eighth day of July, One Thousand Nine Hundred and Fifty-one, in a single copy, of which the English and French texts are equally authentic and which shall remain deposited in the archives of the United Nations, and certified true copies of which shall be delivered to all Members of the United Nations and to the non-member States referred to in article 39.

Note

1   Published by the Office of the UN High Commissioner for Refugees, 1966, HCR/INF/29/ Rev.1

Appendix 3: Protocol Relating to the Status of Refugees, 1967

The States Parties to the present Protocol,

Considering that the Convention relating to the Status of Refugees done at Geneva on 28th July 1951 (hereinafter referred to as the Convention) covers only those persons who have become refugees as a result of events occurring before 1st January, 1951.

Considering that new refugee situations have arisen since the Convention was adopted and that the refugees concerned may therefore not fall within the scope of the Convention.

Considering that it is desirable that equal status should be enjoyed by all refugees covered by the definition in the Convention irrespective of the dateline 1st January 1951,

Have agreed as follows:

Article I
GENERAL PROVISION

1.   The States Parties to the present Protocol undertake to apply articles 2 to 34 inclusive of the Convention to refugees as hereinafter defined.

2.   For the purpose of the present Protocol, the term 'refugee' shall, except as regards the application of paragraph 3 of this article, mean any person within the definition of article 1 of the Convention as if the words 'as a result of events occurring before 1st January 1951 and ...' and the words '... as a result of such events', in article 1A (2) were omitted.

3.   The present Protocol shall be applied by the States Parties hereto without any geographic limitation, save that existing declarations made by States already Parties to the Convention in accordance with Article IB (1) (a) of the Convention shall, unless extended under article 1B (2) thereof, apply also under the present Protocol.

Article II
CO-OPERATION OF THE NATIONAL AUTHORITIES WITH THE UNITED NATIONS

1.   The States Parties to the present Protocol undertake to co-operate with the Office of the United Nations High Commissioner for Refugees, or any other agency of the United Nations which may succeed it, in the exercise of its functions, and shall in particular facilitate its duty of supervising the application of the provisions of the present Protocol.

2.   In order to enable the Office of the High Commissioner, or any other agency of the United Nations which may succeed it, to make reports to the competent organs of the United Nations, the States Parties to the present Protocol undertake to provide them with the information and statistical data requested in the appropriate form, concerning:

(a) The condition of refugees;
(b) The implementation of the present Protocol;
(c) Laws, regulations and decrees which are, or may hereafter be, in force relating to refugees.

Article III
INFORMATION ON NATIONAL LEGISLATION

The States Parties to the present Protocol shall communicate to the Secretary-General of the United Nations the laws and regulations which they may adopt to ensure the application of the present Protocol.

Article IV
SETTLEMENT OF DISPUTES

Any dispute between States Parties to the present Protocol which relates to its interpretation or application and which cannot be settled by other means shall be referred to the International Court of Justice at the request of any one of the parties to the dispute.

Article V
ACCESSION

The present Protocol shall be opened for accession on behalf of all States Parties to the Convention and of any other state member of the United Nations or member of any of the specialized agencies or to which an invitation to accede may have been addressed by the General Assembly of the United Nations. Accession shall be effected by the deposit of an instrument of accession with the Secretary-General of the United Nations.

Article VI
FEDERAL CLAUSE

In the case of a Federal or non-unitary State, the following provisions shall apply:

a) With respect to these articles of the Convention to be applied in accordance with Article 1, paragraph 1, or the present Protocol that come within the legislative jurisdiction of the federal legislative authority, the obligations of the Federal Government shall to this extent be the same as those of States Parties which are not Federal States;

b) With respect to those articles of the Convention to be applied in accordance with Article 1, paragraph 1, of the present Protocol that come within the legislative jurisdiction of constituent States, provinces or cantons which are not, under the constitutional system of the federation, bound to take legislative action, the Federal Government shall bring such articles with a favourable recommendation tot he notice of the appropriate authorities of States, provinces or cantons at the earliest possible moment;

c) A Federal State Party to the present Protocol shall, at the request of any other State Party hereto transmitted through the Secretary-General of the United Nations, supply a statement of the law and practice of the Federation and its constituent units in regard to any particular provision of the Convention to be applied in accordance with Article I, paragraph 1, of the present Protocol, showing the extent to which effect has been given to that provision by legislative or other action.

Article VII
RESERVATIONS AND DECLARATIONS

1.   At the time of accession, any State may make reservations in respect of Article IV of the present Protocol and in respect of the application in accordance with Article I of the present Protocol of any provisions of the Convention other than those contained in Articles 1, 3, 4, 16 (1) and 33 thereof, provided

that in the case of a State Party to the Convention reservations made under this Article shall not extent to refugees in respect of whom the Convention applies.

2.   Reservations made by States Parties to the Convention in accordance with Article 42 thereof shall, unless withdrawn, be applicable in relation to their obligations under the present Protocol.

3.   Any State making a reservation in accordance with paragraph 1 of this Article may at any time withdraw such reservation by a communication to that effect addressed to the Secretary-General of the United Nations.

4.   Declarations made under Article 40, paragraphs 1 and 2, of the Convention by a State Party thereto which accedes to the present Protocol shall be deemed to apply in respect of the present Protocol, unless upon accession a notification to the contrary is addressed by the State Party concerned to the Secretary-General of the United Nations. The Provisions of Article 40, paragraphs 2 and 3, and of Article 44, paragraph 3, of the Convention shall be deemed to apply mutatis mutandis to the present Protocol.

Article VIII
ENTRY INTO FORCE

1.   The present Protocol shall come into force on the day of deposit of the sixth instrument of accession.

2.   For each State acceding to the Protocol after the deposit of the sixth instrument of accession, the Protocol shall come into force on the date of deposit by such State of its instrument of accession.

Article IX
DENUNCIATION

1.   Any State Party hereto may denounce this Protocol at any time by a notification addressed to the Secretary-General of the United Nations.

2.   Such denunciations shall take effect for the State Party concerned one year from the date on which it is received by the Secretary-General of the United Nations.

Article X
NOTIFICATION BY THE SECRETARY-GENERAL OF THE UNITED NATIONS

The Secretary-General of the United Nations shall inform the States referred to in Article V above of the date of entry into force, accessions, reservations and withdrawals of reservations to and denunciations of the present Protocol, and of declarations and notifications relating hereto.

Article XI
DEPOSIT IN THE ARCHIVES OF THE SECRETARIAT OF THE UNITED NATIONS

A copy of the present Protocol, of which the Chinese, English, French, Russian and Spanish texts are equally authentic, signed by the President of the General Assembly and by the Secretary-General of the United Nations, shall be deposited in the archives of the Secretariat of the United Nations. The Secretary-General will transmit certified copies thereof to all States Members of the United Nations and to the other States referred to in Article V above.

Appendix 4: OAU Convention Governing the Specific Aspects of Refugee Problems in Africa, 1969

PREAMBLE

We, the Heads of State and Government assembled in the city of Addis Ababa from 6th to 10th September 1969,

1.   Noting with concern the constantly increasing numbers of refugees in Africa and desirous of finding ways and means of alleviating their misery and suffering as well as providing them with a better life and future,

2.   Recognizing the need for and essentially humanitarian approach towards solving the problems of refugees,

3.   Aware, however, that refugee problems are a source of friction among many Member States, and desirous of eliminating the source of such discord,

4.   Anxious to make a distinction between a refugee who seeks a peaceful and normal life and a person fleeing his country for the sole purpose of fomenting subversion from outside,

5.   Determined that the activities of such subversive elements should be discouraged, in accordance with the Declaration of the Problem of Subversion and Resolution on the Problem of Refugees adopted at Accra in 1965,

6.   Bearing in mind that the Charter of the United Nations and the Universal Declaration of Human Rights have affirmed the principle that human beings shall enjoy fundamental rights and freedoms without discrimination,

7.   Recalling Resolution 2312 (XXII) of 14th December 1967 of the United Nations General Assembly, relating to the Declaration on Territorial Asylum,

8.   Convinced that all the problems of our continent must be solved in the spirit of the Charter of the Organization of African Unity and in the African context,

9.   Recognizing that the United Nations Convention of 28th July 1951, as modified by the Protocol of 31st January 1967, constitutes the basic and universal instrument relating to the status of refugees and reflects the deep

concern of States for refugees and their desire to establish common standards for their treatment,

10.   Recalling Resolutions 26 and 104 of the OAU Assemblies of Heads of State and Government, calling upon Member States of the Organization who had not already done so to accede to the United Nations Convention of 1951 and to the Protocol of 1967 relating to the Status of Refugees, and meanwhile to apply their provisions to refugees in Africa,

11.   Convinced that the efficiency of the measures recommended by the present Convention to solve the problem of refugees in Africa necessitates close and continuous collaboration between the Organization of African Unity and the office of the United Nations High Commissioner Refugees, have agreed as follows:

Article I
DEFINITION OF THE TERM 'REFUGEE'

1.   For the purposes of this Convention, the term 'refugee' shall mean every person who, owing to well-founded fear of being persecuted for reasons of race, religion, nationality, membership of a particular social group of political opinion, is outside the country of his nationality and is unable or, owing to such fear, is unwilling to avail himself of the protection of that country, or who, not having a nationality and being outside the country of his former habitual residence as a result of such events is unable or, owing to such fear, is unwilling to return to it.

2.   The term 'refugee' shall also apply to every person who, owing to external aggression, occupation, foreign domination or events seriously disturbing public order in either part or the whole of his country of origin or nationality, is compelled to leave his place of habitual residence in order to seek refuge in another place outside his country of origin or nationality.

3.   In the case of a person who has several nationalities, the term 'a country of which is his a national' shall mean each of the countries of which he is a national, and a person shall not be deemed to be lacking the protection of the country of which he is a national if, without any valid reason based on well-founded fear, he has not availed himself of the protection of one of the countries of which he is a national.

4.   This Convention shall cease to apply to any refugee if:

a) he has voluntarily re-availed himself of the protection of the country of his nationality, or

b) having lost his nationality, he voluntarily re-acquired it, or

c) he has acquired a new nationality, and enjoys the protection of the country of his new nationality, or

d) he has voluntarily re-established himself in the country which he left or outside which he remained owing to fear of persecution, or

e) he can no longer, because the circumstances in connection with which he was recognized as a refugee have ceased to exist, continue to refuse to avail himself of the protection of the country of his nationality, or

f) he has committed a serious non-political crime outside his country of refuge after his admission to that country as a refugee, or

g) he has seriously infringed the purposes and objectives of this Convention.

5.   The provisions of this Convention shall not apply to any person with respect to whom the country of asylum has serious reasons for considering that:

a) he has committed a crime against peace, a war crime, or a crime against humanity, as defined in the international instruments drawn up to make provision in respect of such crimes;

b) he committed a serious non-political crime outside the country of refuge prior to his admission to that country as a refugee;

c) he has been guilty of acts contrary to the purposes and principles of the Organization of African Unity;

d) he had been guilty of acts contrary to the purposes and principles of the United Nations.

6.   For the purposes of this Convention, the Contracting State of asylum determine whether an applicant is a refugee.

Article II
ASYLUM

1.   Member States of the OAU shall use their best endeavours consistent with their respective legislations to receive refugees and to secure the settlement of those refugees who, for well-founded reasons, are unable or unwilling to return to their country of origin or nationality.

2.   The grant of asylum to refugees is a peaceful and humanitarian act and shall not be regarded as an unfriendly act by any Member State.

3.   No person shall be subjected by a Member State to measures such as rejection at the frontier, return or expulsion, which would compel him to return to or remain in a territory where his life, physical integrity or liberty would be threatened for the reasons set out in Article I, paragraphs 1 and 2.

4.   Where a Member State finds difficulty in continuing to grant asylum to refugees, such Member State may appeal directly to other Member States and through the OAU, and such other Member States shall in the spirit of African solidarity and international co-operation take appropriate measures to lighten the burden of the Member State granting asylum.

5.   Where a refugee has not received the right to reside in any country of asylum, he may be granted temporary residence in any country of asylum in which he first presented himself as a refugee pending arrangement for his resettlement in accordance with the proceeding paragraph.

6.   For reasons of security, countries of asylum shall, as far as possible, settle refugees at a reasonable distance from the frontier of their country of origin.

Article III
PROHIBITION OF SUBVERSIVE ACTIVITIES

1.   Every refugee has duties to the country in which he finds himself, which require in particular that he conforms with its laws and regulations as well as with measures taken for the maintenance of public order. He shall also abstain from any subversive activities against any Member State of the OAU.

2.   Signatory States undertake to prohibit refugees residing in their respective territories from attacking any State Member of the OAU, by any activity likely to cause tension between member states, and in particular by use of arms, through the press, or by radio.

Article IV
NON-DISCRIMINATION

Member states undertake to apply the provisions of this Convention to all refugees without discrimination as to race, religion, nationality, membership of a particular social group or political opinions.

Article V
VOLUNTARY REPATRIATION

1.   The essentially voluntary character of repatriation shall be respected in all cases and no refugee shall be repatriated against his will.

2.   The country of asylum, in collaboration with the country of origin, shall make adequate arrangements for the safe return of the refugees who request repatriation.

3.   The country of origin, on receiving back refugees, shall facilitate their resettlement and grant them the full rights and privileges of nationals of the country, and subject them to the same obligations.

4.   Refugees who voluntarily return to their country shall in no way be penalized for having left it for any of the reasons giving rise to refugee situations. Whenever necessary, an appeal shall be made through national information media and through the Administrative Secretary-General of the OAU, inviting refugees to return home and giving assurance that the new circumstances prevailing in their country of origin will enable them to return without risk and to take up a normal peaceful life without fear of being disturbed or punished, and that the text of such appeal should be given to refugees and clearly explained to them by their country of asylum.

5.   Refugees who freely decide to return to their homeland, as a result of such assurances or on their own initiative, shall be given every possible assistance by the country of asylum, the country of origin, voluntary agencies and international and intergovernmental organizations, to facilitate their return.

Article VI
TRAVEL DOCUMENTS

1.   Subject to Article III, Member States shall issue to refugees lawfully staying in their territories travel documents in accordance with the United Nations Convention relating to the Status of Refugees and the Schedule and Annex thereto, for the purpose of travel outside their territory, unless compelling reasons of national security of public order otherwise require, Member States may issue such a travel document to any other refugee in their territory.

2.   Where an African country of second asylum accepts a refugee from a country of first asylum, the country of first asylum may be dispensed from issuing a document with a return clause.

3.   Travel documents issued to refugees under previous international agreements by States Parties thereto shall be recognized and treated by Member States in the same way as if they had been issued to refugees pursuant to this Article.

Article VII
CO-OPERATION OF THE NATIONAL AUTHORITIES WITH THE ORGANIZATION OF AFRICAN UNITY

In order to enable the Administrative Secretary-General of the Organization of African Unity to make reports to the competent organs of the Organization of African Unity, Member States undertake to provide the Secretariat in the appropriate form with information and statistical data requested concerning:
a) the condition of refugees;
b) the implementation of this Convention, and
c) laws, regulations and decrees which are, or may hereafter be, in force relating to refugees.

Article VIII
CO-OPERATION WITH THE OFFICE OF THE UNITED NATIONS HIGH COMMISSIONER FOR REFUGEES

1.   Member States shall co-operate with the office of the United Nations High Commissioner for Refugees.
2.   The present Convention shall be the effective regional complement in Africa of the 1951 United Nations Convention on the Status of Refugees.

Article IX
SETTLEMENT OF DISPUTES

Any dispute between States signatories to this Convention relating to its interpretation or application, which cannot be settled by other means, shall be referred to the Commission for Mediation, Conciliation and Arbitration of the Organization of African Unity, at the request of any one of the Parties to the dispute.

Article X
SIGNATURE AND RATIFICATION

1.   This Convention is open for signature and accession by all Member States of the Organization of African Unity and shall be ratified by signatory States in accordance with their respective constitutional processes. The instruments or ratification shall be deposited with the Administrative Secretary-General of the Organization of African Unity.

2. The original instrument, done if possible in African languages, and in English and French, all texts being equally authentic, shall be deposited with the Administrative Secretary-General of the Organization of African Unity.

3. Any independent African State, Member of the Organization of African Unity, may at any time notify the Administrative Secretary-General of the Organization of African Unity of its accession to this Convention.

Article XI
ENTRY INTO FORCE

This Convention shall come into force upon deposit of instruments of ratification by one-third of the Member States of the Organization of African Unity.

Article XII
AMENDMENT

This Convention may be amended or revised if any Member State makes a written request to the Administrative Secretary-General to that effect, provided however that the proposed amendment shall not be submitted to the Assembly of Heads of State and Government for consideration until all Member States have been duly notified of it and a period of one year has elapsed. Such an amendment shall not be effective unless approved by at least two-thirds of the Member States Parties to the present Convention.

Article XIII
DENUNCIATION

1. Any Member State Party to this Convention may denounce its provisions by a written notification to the Administrative Secretary-General.

2. At the end of one year from the date of such notification, if not withdrawn, the Convention shall cease to apply with respect to the denouncing state.

Article XIV

Upon entry into force of this Convention, the Administrative Secretary-General of the OAU shall register it with the Secretary-General of the United Nations, in accordance with Article 102 of the Charter of the United Nations.

Article XV
NOTIFICATIONS BY THE ADMINISTRATIVE SECRETARY-GENERAL OF THE ORGANIZATION OF AFRICAN UNITY

The Administrative Secretary-General of the Organization of African Unity shall inform all Members of the Organization:
a) of signatures, ratifications and accessions in accordance with Article X;
b) of entry into force, in accordance with Article XI;
c) of requests for amendments submitted under the terms of Article XII;
d) of denunciations, in accordance with Article XIII.

Appendix 5: Refugees (Control) Act, 1965 of the United Republic of Tanzania

THE UNITED REPUBLIC OF TANZANIA
No. 2 of 1966

I ASSENT,
J.K. NYERERE,
President

6th January, 1966

An Act to make provision for the Control of Refugees and for connected matters

[7th JANUARY, 1966]

ENACTED by the Parliament of the United Republic of Tanzania.

Short Title.
1.   This Act may be cited as the Refugee (Control) Act, 1965.

Interpretation.
2.   In this Act, unless the context otherwise requires – 'area' in relation to a competent authority means–
   (a) when the competent authority is a regional commissioner, the region of which he has the charge; and
   (b) when the competent authority is an area commissioner, the district of which he has the charge;
   'authorized officer' means an administrative officer, a settlement commandant, a policy officer, a prisons officer or a member of the Tanzania People's Defence Forces;

'competent authority' means a regional commissioner and, for the purposes of sections 5, 6, 7, and 8 includes an area commissioner;

'the Minister' means the Minister for the time being responsible for refugees;

'reception area' means an area declared as such by the Minister under section 4;

'refugee' means one of a class of persons declared to be refugees by the Minister under section 3 other than a person to whom subsection (2) of that section refers;

'refugee' settlement' means a refugee settlement established in accordance with section 4;

'settlement commandant' means a person appointed to be in charge of a refugee settlement.

Refugees.

3.– (1) Subject to the proviisons of subsection (2), the Minister may –

    (a) by order published in the Gazette, declare any class of persons who are, or prior to their entry into Tanganyika were, ordinarily resident outside Tanzania to be refugees for the purposes of this Act;

    (b) by the same or any subsequent such order declare that the provisions of sections 11 and 12, or of either of them, shall apply to refugees or to any category of refugees.

  (2) No declaration made under paragraph (1) of subsection (1) shall apply to –

    (a) any citizen of Tanzania;

    (b) any person entitled in Tanzania to diplomatic immunity;

    (c) any person in the employment of any state, government or local authority outside Tanzania, or of any international organization for the time being specified in the Third Schedule to the Immunities and Privileges (Extension and Miscellaneous Provisions) Ordinance, who enters Tanzania in the course of his duties;

    (d) any member of a class of persons declared by the Minister, by order published in the Gazette, not to be refugees for the purposes of this Act.

  (3) If any question arises in any proceedings, or with reference to anything done or proposed to be done, under this Act as to whether any person is a refugee or not, or is a refugee of a particular category or not, the

onus of proving that such person is not a refugee or, as the case may be, is not a refugee of a particular category, shall lie upon that person.

Reception areas and refugee settlements.
4.– (1) The Minister may declare any part of Tanganyika to be an area for the reception or residence of any refugees or category thereof.

(2) The competent authority may establish in any reception area a refugee settlement for refugees or any category thereof, and may appoint a settlement commandant to be in charge of such settlement.

Provisions applying to refugees generally. Places of entry or departure and routes.
5.– (1) The Minister or, as respects his area, the competent authority may, by order in writing –
   (a)   direct that any refugee entering or leaving Tanganyika shall enter or leave by specified routes or at specified places;
   (b)   direct that any refugee moving from one part of Tanganyika to another shall move by specified routes.

(2) Orders made under this section may be subject to such terms and conditions as the Minister or, as the case may be, the competent authority may think fit.

(3) Any refugee who contravenes an order made under this section or the terms or conditions therof shall be guilty of an offence against this Act.

Surrender of weapons.
6.– (1) Every refugee who brings any arms or ammunition into Tanganyika shall immediatelysurrender such arms or ammunition to an authorized officer.

(2) The competent authority may, by order in writing, direct that any refugee in his area shall, within such time as may be specified in the order, surrender to an authorized officer any other weapon or weapons, or any instrument or tool so specified which is capable of being used as a weapon and which is in, or comes into, his possession unless the possessor thereof has written authority to retain the same signed by the competent authority or an authorized officer appointed by the competent authority in that behalf.

(3) Any refugee who fails to surrender any arms, ammunition, weapon, instrument or tool in accordance with this section or any order made

hereunder shall be guilty of an offence and shall be liable on conviction to imprisonment for a term not exceeding two years.

(4) In subsection (1) of this section, 'arms' and 'ammunition' have the meanings respectively ascribed to those expressions in the Arms and Ammunition Ordinance, and the provisions of this section in relation to arms and ammunition are in addition to, and not in subsitution for the provisions of that Ordinance.

Detention and slaughter of animals.

7.– (1) The competent authority may direct that any animal imported into his area from outside Tanganyika by any person whom he has reason to believe to be a refugee shall be kept in such place as he shall direct or shall be slaughtered or otherwise disposed of.

(2) If any animal is slaughtered or sold as a result of any direction given under the provisions of subsection (1), the competent authority shall use his best endeavours to ensure that the person owning the animal sold shall be paid the proceeds of the sale less the expenses of the sale.

(3) The proceeds of a sale directed under the provisions of subsection (1), less the expenses of the sale, shall, if they are not paid to the owner of the animal, be apid into a fund which shall be used for the benefit of refugees.

(4) Notwithstanding any other provision of this section, if a veterinary officer in the service of the Government is of the opinion that in order to prevent the spread of disease it is necessary to slaughter any animals which he has reason to believe belong to refugees such animals shall be slaughtered as the veterinary officer directs.

(5) Any person who in any way obstructs the carrying out of any direction given under the provisions of this section shall be guilty of an offence against this Act.

Detention and use of vehicles.

8.   The competent authority may take, or authorize an authorized officer to take, possession of any vehicle in which any person whom he has reasonable cause to believe to be a refugee arrives in his area and may authorize its use in that area for the purpose of moving refugees or any stores or equipment for their use.

Deportation of Refugees.

9.– (1) The Minister or, as respects his area, any competent authority appointed by the Minister in that behalf may at any time order any refugee to return by such means or route as he shall direct to the territory from which he entered Tanganyika.

(2) A court convicting any refugee of an offence under the provisions of this section mayorder the deprtation of such refugee tothe territory from which he entered Tanganyika.

(3) Where any person is ordered to return tothe territory from which he entered Tanganyika or to be deported under subsection (1) or (2) he may be held in custody and deported in accordance with such order.

(4) No order shall be made under subsection (1) or (2) in respect of a refugee if the Minister, the competent authority or the court, as the case may be, is of the opinion that such a refugee will be tried or punished for an offence of a political character after arrival in the territory from which he came or is likely to be the subject of physical attack in such territory.

(5) Any refugee failing to comply with an order made under subsection (1) shall be guilty of an offence against this Act.

(6) Where an order is made under this section in respect of a refugee who has been present in Tanzania for a continuous period of not less than three moths immediately prior to the making of the order, the authority making the order shall inform the refugee, or cause him to be informed, that he may make representations against his deportation on the grounds that he is in danger ofbeing tried or punished for an offence of a political character after arrival in the territory from which he came or is in danger of physical attack in such territory. A refugee to whom this subsection applies who wishes to make such representations shall make them forthwith to the person by whom he is so informed and that person shall reduce such representations to writing and forward them to the Minister; and the Minister shall consider the same and determine whether or not the refugee shall be deprted in accordance with the order in that behalf or whether that order shall be revoked, and where the Minister determines that the order shall be revoked, he shall have power to revoke the same. Pending the determination of the Minister on anysuch representations, the order for the deportation of the refugee shall be suspended.

Detention of refugees who prejudice peace, order or foreign relations, or who are believed to have committed offences outside Tanzania.

10.–(1) If the Minister or, as respects his area, any competent authority appointed by the Minister in that behalf is satisfied that any refugee is acting in a manner prejudicial to peace and good order or is prejudicing the relations between the Government of Tanzania and any other Government, he may, by order in writing, direct that the refugee be detained in prison.

(2) If it appears to the Minister or, as respects his area, any competent authority appointed by the Minister in that behalf that it is likely that any refugee has committed any offence in any other territory for which he has not been punished, being an offence which, if committed within Thanganyika, would be punishable by imprisonment, the Minister or such competent authority may, by order in writing, direct that such refugee be detained in prison.

(3) Any order under subsection (1) or (2) shall be sufficient authority for any authorized officer to arrest the refugee to whom it applies and to detain him in custody pending or during his transportation to prison and for the officer in charge of a prison to hold such refugee in custody as an unconvicted prisoner until his release is ordered by the Minister:

Provided that where any such order is made by a competent authority the order shall, unless previously confirmed by the Minister, expire at the end of the fourteenth day after that on which the refugee was arrested.

(4) Where, in accordance with the proviso to subsection (3), any order expires on account of its not having been confirmed by the Minister, the refugee to whom the order applies shall not be arested again by order of the competent authority for the same cause.

(5) The imprisonment of every refugee in accordance with this section shall be reviewed by the Minister at intervals of not less than three months or such lesser period as the Minister shall direct.

Special provisions which may be applied to refugees. Permits to remain in Tanganyika.

11.–(1) No refugee to whom this section applies shall remain in Tanganyika –

a) unless within seven days of his entering Tanganyika or, if at the time of his entry this section does not apply to him, within seven days after its being applied to him, he is issued with a permit to remain by an authorized officer appointed by the competent authority in that behalf;

b) unless he complies with the terms or conditions from time to time annexed to such permit by the competent authority.

(2) An authorized officer appointed in that behalf shall not refuse a refugee a permit under this section if the officer has reason to believe that the refusal of a permit will necesssitate the return of the refugee to the territory from which he entered Tanganyika and that the refugee will be tried or punished for an offence of a political character after arrival in that territory or is likely to be the subject of physical attack in that territory; but, save as aforesaid, such authorized officer may in his discretion and without assigning any reason refuse to issue a permit.

(3) If a refugee to whom this section applies fails to obtain or is refused a permit in accordance with this section, his presence in Tanganyika shall, notwithstanding anything contained in section 2 of the Immigration Act, 1963, be unlawful.

(4) Where a permit is refused under this section in respect of a refugee who has been present in Tanzania for a continuous period ofnot less than three months immediately prior to the refusal, the authorized officer shall inform the refugee that he may make representations against such refusal on the grounds that the refusal of the permit will necessitate the return of the refugee to the territory from which he entered Tanganyika and that he is in danger of being tried or punished for an offence of a political character after arrival in that territory or is in fanger of physical attack in that territory. A refugee to whom this subsection applies who wishes to make such representations shall make them forthwith to the authorized officer and that officer shall reduce such representations to writing and forward them to the Minister; and the Minister shall consider the same and determine whether or not a permit shall be granted to the refugee, and where the Minister determines that a permit shall be granted, the authorized officer shall grant a permit accordingly. Pending the determination of the Minister on any such representations, subsection (3) shall not have effect with respect to the refugee.

Requirement to reside in reception area or refugee settlement.

12.–(1) The competent authority may –

a) by order, require any refugee to whom this sectoin applies who is within his area to reside within a reception area or refugee settlement, whether such reception area or refugee settlement is within such competent authority's area or not;

b)   require any refugee to whom this section applies who is within a reception area or refugee settlement within such competent authority's area to remove to and reside in some other place being a reception area or refugee settlement, whether such other place is within such competnet authority's area or not.

(2) Any refugee to whom an order made under this section applies who –

a)   fails to take steps forthwith to comply with such order; or

b)   fails to move to or take up residence in a reception area or refugee settlement in accordance with such order with reasonable despatch; or

c)   having arrived at a reception area or a refugee settlement in pursuance of such order, leaves or attempts to leave such area or settlement except in pursuance of some other order made under this section, shall, unless he is in possession of a permit issued in thatbehalf under subsection (3), be guilty of an offence against this Act.

(3) The competent authority, or an authorized officer appointed by a competent authority in that behalf, may issue a permit to any refugee to whom an order made under subsection (1) applies, authorizing him –

a)   to reside in a reception area elsewhere than in the refugee settlement to which such order refers;

b)   to leave a reception area in which he has been required to reside.

(4) The competent authority or such authorized officer may issue a permit under this section subject to such terms and conditions as he thinks fit, and, without prejudice to the generality of the foregoing he may, where he issues a permit under paragraph (b) or subsection (3), specify the destination to which and the route by which such refugee may proceed.

(5) Any refugee to whom a permit has been issued under this section who fails to comply with the terms and conditions thereof, shall be guilty of an offence against this Act.

Control of Refugee Settlements.

13.–(1) The Minister may make rules, and the competent authority may issue directions not inconsistent with such rules, for the control of refugee settlements and, without prejudice to the generality of the foregoing, such rules and directions may make provision in respect of all or any of the following matters –

a) The organization, safety, discipline and administration of such settlements;

b) the reception, treatment, health and well-being of refugees;

c) the manner of inquiring into disciplinary offences and the payment of fines;

d) the establishment and regulation of settlement lock-ups and the custody of persons therein; and

e) the powers of settlement commandants and the delegation of such powers.

(2) A settlement commandant may give such orders or directions, either orally or in writing, to any refugee as may be necessary or expedient for the following purposes, that is to say –

a) to ensure that the settlement is administered in an orderly and efficient manner;

b) to ensure the performance of any work or duty necessary for the maintenance of essential services in the settlement or for the general welfare of the refugees therein;

c) to ensure that all proper precautions are taken to preserve the health and well-bing of the reguees therein;

d) to preserve orderly conduct and discipline in the settlement.

(3) Any refugee who –

a) without a permit in that behalf issued under section 12, leaves or attempts to leave a refugee settlement in which he has been ordered to reside; or

b) in a refugee settlement disobeys any rules made by the Minister, any direction of the competent authority, or any order or direction of a settlement commandant, made or given under this section; or

c) in a refugee settlement conducts himself in a manner prejudicial to good order and discipline, shall be deemed to have committed a disciplinary offence.

(4) The competent authority or a settlement commandant may inquire into any disciplinary offence and if he finds that a refugee has been guilty of such disciplinary offence may punish him by –

(a) ordering his confinement in a settlement lock-up for a period not exceeding thirty days; or

(b) fining him a sum not exceeding Shs.200/–.

(5) Any refugee convicted of a disciplinary offence and sentenced by a settlement commandant to confinement for a period exceeding fourteen days or to a fine exceeding Shs.100/– amy appeal to the competent authority whose decision shall be final.

Miscellaneous

Restrictions on Persons entering refugee settlements or addressing refugees in settlements.

14.–(1) No person other than a refugee required to reside or residing in, or a person employed in, a refugee settlement, shall enter or be within such settlement except with the general or special permission of the Minister, the competent authority or the settlement commandant.

(2) No person other than the competent authority, an authorized officer or a person authorized in that behalf by the competent authority or the settlement commandant, may in a refugee settlement address an assembly or meeting of more than ten refugees whether or not such meeting is held in a public place.

(3) Any person who contravenes the provisions os this section shall be guilty of an offence against this Act.

Offences and Penalties.

15.–(1) Any refugee who fails to obey any lawful order of the Minister, the competent authority or of a settlement commandant or who obstructs a competent authority or authorized officer in the exercise of his powers under this Act, shall be guilty of an offence and shall be liable on conviction to imprisonment for a period not exceeding three months.

(2) Any person who is guilty of an offence against this Act for which no penalty is specifically provided shall be liable on conviction to imprisonment for a period not exceeding three months.

3) Where any act or omission constitutes both a disciplinary offence under section 13 and an offence punishable under this Act on conviction, a refugee shall not be punished for the same act or omission both as a disciplinary offence and as an offence punishable on conviction.

Arrest.

16.   An authorized officer may arrest without warrant any person whom he has reasonable grounds for suspecting to have committed an offence or a disciplinary offence under this Act and such person may be detained in custody

in a settlement lock-up or in any prison or police station pending the institution of proceedings for the offence or disciplinary offence.

Force.
17.   An authorized officer or any person acting with the authority of an authorized office may use such force (including the use of firearms) as may be necessary to compel any refugee to comply with any order or direction, whether oral or in writing, given pursuant to the provisions of the Act.

Publications of, and powers in relation to orders, etc.
18.–(1) Any rules, orders or directions under this Act may be published in such manner as the authority making the same considers appropriate in order to bring the same to the notice of the persons to whom they apply or who are affected thereby but, save as may be expressly required by the Act, shall not be required to be published in the Gazette:
    Provided that any such order or direction which is applied or directed to any individual person shall, if in writing, be served on, and if not in writing, be given to, the person affected personally.
(2) Where under sections 5, 6, 7, 8, 9, 11, 12, or 13 orders or directions may be given or applied to, or given or made in respect of, refugees or a refugee, such rules, directions or orders may be given or applied to, or given or made in respect of, all refugees, or any particular refugees or category of refugees.

Protection of bonafide act.
19.–(1) No act or thing done or omitted to be done by any person shall, if the act or omission was done or omitted bonafide while acting in the execution of his duty under this Act, subject him personally to any liability, action, claim or demand whatever.
(2) Save as provided in subjection (2) of section 7, no compensation shall be payable and no action shall be brought against any person acting in the execution of his duty under this Act for any damage done or loss occasioned by, the detention or slaughter of any animal under the powers conferred by section 7, or the detention or use of any vehicle under the powers conferred by section 8.

Powers of competent authorities outside their areas.
20. Where a competent authority having power in that behalf –
    (a)   orders any refugee to be detained in prison under section 10; or

    (b)   orders any refugee to reside in any reception area or refugee settlement under section 12, the order shall have effect and be given effect to notwithstanding that the prison in which the refugee is to be, or is, detained, or that the reception area or refugee settlement in which the refugee is to reside or resides, is outside the oarea of such competent authority.

21.   The War Refugees (Control and Expulsion) Ordinance is hereby repealed.

Passed in the National Assembly on the twenty-third day of December, 1965.

P. MESKWA.
Clerk of the National Assembly

Appendix 6: The Refugees (Recognition and Control) Act, 1967 of Botswana

GOVERNMENT OF BOTSWANA
THE REFUGEES (RECOGNITION AND CONTROL) ACT, 1967
(No. 9 of 1967)
(Promulgated 28th April, 1967)
as amended by
THE REFUGEES (RECOGNITION AND CONTROL) (AMENDMENT)
ACT, 1967
(No. 37 of 1967)
(Promulgated 13th October, 1967)

Reprinted by direction of the Attorney-General in terms of section 3 of the Amendments Incorporation Law, 1966.

Date of reprint:

ARRANGEMENT OF SECTIONS

THE REFUGEES (RECOGNITION AND CONTROL) ACT, 1967
(No. 8 of 1967)

AN ACT TO MAKE PROVISION FOR THE RECOGNITION AND CONTROL OF CERTAIN POLITICAL REFUGEES: TO PREVENT IN CERTAIN CIRCUMSTANCES THEIR REMOVAL FROM BOTSWANA UNDER THE IMMIGRATION (CONSOLIDATION) LAW, 1966; AND TO MAKE PROVISION INCIDENTAL THERETO OR CONNECT THEREWITH.

Date of Assent: 21.4.67.
Date of Commencement: 5.5.67.
ENACTED by the Parliament of Botswana.

PART 1
PRELIMINARY

Short Title and Commencement.
1.   This Act may be cited as the Refugees (Recognition and Control) Act, 1967, and shall come into operation on a date to be appointed by the Minister by notice in the Gazette.

Interpretation.
2.   (1) In this Act, unless the context otherwise requires – 'Committee' means a Refugee Advisory Committee established under section 3; 'Convention'[1] means the United Nations Convention Relating to the Status of Refugees of the 28th July, 1951, as amended from time to time by any international agreement; but does not include any provisions thereof not binding under public international law upon the Republic of Botswana;
'immigrant' means any person in Botswana other than –

(a)   a citizen of Botswana; or

(b)   an established resident;

'Immigration Law' means the Immigration (Consolidation) Law, 1966 (No. 19 of 1966);

'political refugee'[2] means an immigrant whom the Minister has declared in terms of section 8 (1) that he recognizes as a political refugee;

'removed from Botswana' does not include deportation in terms of section 24 of the Immigration Law.

(2) Subject to the provisions of subsection (1) and unless the context otherwise requires, any word or expression defined in the Immigration Law shall bear the same meaning in this Act as in the Immigration Law.

Establishment of Refugee Advisory Committees.

3.   (1) The Minister may, by notice in the Gazette, establish one or more Refugee Advisory Committees to carry out the functions conferred on such Committees by or under this Act.

(2) A committee shall consist of a Chairman and not less than two, nor more than four, other members.

PART II
REFUGEES

Enquiry by committee.[3]

4.   (1) Unless the Minister otherwise directs, a committee shall hold an enquiry into the case of any immigrant who on presenting himself to an immigration officer in terms of section 6 of the Immigration Law claims to be a political refugee.

(2) A Committee shall also hold an enquiry into the case of any other immigrant who in the opinion of the Minister is in Botswana in such circumstances as indicate may be a political refugee.

(3) After holding an enquiry in terms of this section a committee shall report thereon to the Minister.

Powers and Procedures of Committee.

5.   (1) For the purpose of conducting an enquiry in terms of section 4, a committee shall have power –

(a) by notice under the hand of its Secretary or Chairman, to summon before it any person in respect of whom the enquiry is to be held;

(b) by notice under the hand of its Secretary or Chairman, to summon before it any person who may be able to give information which will assist the Board, or call upon him to submit such information in writing;

(c) to examine any person appearing before it on oath or otherwise;

(d) to call upon any person to furnish the committee with such information as it considers will assist it in the exercise of its functions whether in the form of a statutory declaration, in writing, orally or otherwise and to produce to the committee any documents which are in his possession or under his control and which the committee considers may be relevant to the enquiry.

(2) The proceedings of a committee shall be in private and shall be conducted in such manner as the committee may determine.

Provided that the immigrant who is the subject of the enquiry shall be notified thereof and be given the opportunity of appearing before the committee and of making representations concerning his case to it.[4]

(3) Any person who –

(a) refuses or fails without sufficient reason to appear before a Committee at the time and place specified in a notice given under subsection (1) (a) or (b);

(b) gives false evidence or information to a committee or who attempts to mislead the committee;

(c) fails to comply with a notice given under subsection (1) (d);

shall be guilty of an offence and liable on conviction to a fine not exceeding R500 or to imprisonment for a period not exceeding six months or to both such fine and such imprisonment.

Restriction on Removal of Immigrant Who May be a Refugee.

6.　　　Where an immigrant who is liable to be removed from Botswana under the provisions of the Immigration Law is summoned to appear before a committee under the provisions of section 5 (1) (a) –

(a) he shall not be so removed pending a determination by the Minister in accordance with the provisions of section 8; and

(b) pending such determination he may be detained by an immigration officer for a period not exceeding 28 days; if he is so detained the provisions of section 15 (2) and (3) of the Immigration Law shall apply in relation to him as if he were being detained under subsection (1) of that section.

Right of Detained Immigrant to Leave Botswana.

7.   Notwithstanding the provisions of section 6 (b), any person detained in pursuance of that provision shall, unless liable to detention under some other lawful authority, be allowed to depart from Botswana for the purpose of entering some other country if he satisfies an immigration officer that it is lawful for him to enter such other country without his possessing a right of re-entry to Botswana and that he possesses the means and in fact intends to enter that country.

Recognition of Immigrant as Political Refugee.

8.   (1) When the Minister receives the report of an enquiry held in terms of section 4 he may –
- (a) subject to the provisions of paragraph (b), if he is of the opinion that the person who has been the subject of the enquiry is a political refugee, declare that he recognizes such person as a political refugee; or
- (b) if he is of the opinion that the person who has been the subject of the enquiry is not a political refugee or if he considers that there is no or insufficient reason to treat him as a political refugee declare that he does not recognize such person as a political refugee; or
- (c) direct the Committee to reopen the enquiry or to make further report in the matter.

(2) Where, in terms of subsection (1), the Minister declares that he does not recognize a person as a political refugee such person shall, if liable to be removed from Botswana under the Immigration Law, be so removed and shall, whether so liable or not, be subject in all respects to the provisions of that law.

(3) Save where this Act otherwise provides, a person who is recognized as a political refugee shall be subject to the provisions of the Immigration Law in all respects as if the declaration of recognition had not been made.

Restriction on Removal and Control of Refugee.

9.   (1)[5] Subject to the provision of section 10, a recognized refugee shall not be removed from Botswana under the provisions of the Immigration

Law except to a country approved by the Minister, being a country in which, in the opinion of the Minister the life or freedom of the refugee will not be threatened on account of his race, religion, nationality or membership of a particular social group or political opinion;

Provided that nothing in this subsection shall prevent the removal, under the provisions of any law, of a recognized refugee to any country whatsoever where, in the opinion of the Minister such removal is desirable on the grounds of national security or of public order or where the recognized refugee has been convicted by a final judgement of any court of a serious crime which, in the opinion of the Minister, indicates that the recognized refugee constitutes a danger to the community.

(2) Notwithstanding the provisions of subsection (1), a recognized refugee who is liable to be removed from Botswana under the provisions of the Immigration Law may be detained by an immigration officer pending such removal, and if he is so detained the provisions of section 15 (2) and (3) of the Immigration Law shall apply in relation to him as if he were being detained under subsection (1) of that section.

Provided that where in the opinion of the Minister, delay is likely to occur before such removal may be effected the Minister may, in his sole and absolute discretion, direct that the refugee shall not be detained under this subsection but shall while he remains in Botswana be subject to all or any of the following conditions –

(a)   that the refugee shall reside at a place or within an area specified by the Minister;

(b)   that the refugee shall not depart from such place or area or only depart therefrom subject to such conditions as may be specified by the Minister;

(c)   that the refugee shall give recognizances for his good behavior in such form and subject to such conditions as may be specified by the Minister;

(d)   that the refuge shall report to the police or such other authority as may be specified by the Minister in such manner as he may determine;

(e)   that the refugee shall not take an active part in the politics of Botswana or of any other country in Africa or not take part in such activities, being activities of a political nature, as may be specified by the Minister;

(f) such ancillary or additional conditions as may appear to the Minister to be necessary or desirable in the circumstances of the case.

(3) The Minister may at any time withdraw or modify a direction under the proviso to subsection (2).

(4) Any recognized refugee who having been released from detention in terms of the proviso to subsection (2) fails to comply with any condition of such release shall be guilty of an offence and liable on conviction to a fine not exceeding R500 or to imprisonment for a period not exceeding six months or to both such fine and such imprisonment.

Departure of Refugee from Botswana.

10. (1) A recognized refugee who is not detained under section 9 (2) or other lawful authority may leave Botswana at any time.

(2) A recognized refugee shall on his departure from Botswana cease to be a recognized refugee.

(3) Notwithstanding the provisions of section 9 (1) any recognized refugee who is detained under section 9 (2) shall, unless he is liable to detention under some other lawful authority, be allowed to depart from Botswana for the purpose of entering some country other than a country approved by the Minister in terms of section 9 (1) if he certifies in writing that he wishes to enter that other country and satisfies an immigration officer that it is lawful for him to enter that country without his possessing a right of re-entry to Botswana and that he possesses the means to do so.

Review of Case of Recognized Refugee.

11. (1) Not more than six months after the recognition of a refugee under section 8, and thereafter at intervals of not more than six months, the case of that refugee shall be reviewed by a committee, who shall advise the Minister –

(a) whether to exercise any of his powers under this Act or the Immigration Law in relation to that refugee;

(b) as to the moral and economic welfare of that refugee; and what steps should be taken to secure the same.

(2) On receiving the report of a review held in terms of subsection (1) the Minister may –

(a)   if he considers that there is no or insufficient reason to continue treating the refugee as a political refugee declare that he no longer recognizes him as a political refugee;

(b)   direct the committee to reopen the review or to make further report in the matter;

(c)   take such alternative or additional steps open to him under this Act or otherwise in relation to the refugee as may appear to him most proper.

(3) The provisions of section 5 shall have effect for the purposes of a review under subsection (1) as they have for the purposes of an enquiry under section 4.

Recognizances.

12.   Where under the provisions of the proviso to section 9 (2) a recognized refugee is required to give recognizances and such recognizances include the entering into of a bond for an amount of money to be forfeited if the conditions of the bond are broken, the Chief Immigration Officer may, upon breach of any condition of the bond, make application to a court of competent jurisdiction which may give judgement against the refugee or his sureties in accordance with the conditions of the bond.[6]

Residence of Refugee Not Ordinary Residence.

13.   For the purposes of any other written law, other than a taxation law, any period during which an immigrant has resided in Botswana as a recognized refugee shall not unless the Minister in writing otherwise directs1 be regarded as a period during which he has been ordinarily resident in Botswana.[7]

Regulations.

14. The Minister may make regulations –

(a) providing for the custody of the property of any political refugee who is detained;

(b) prescribing the form of any notice which may be given under this Act;

(c) prescribing the allowances payable to members of a Committee and the fees payable to persons giving evidence before it;

(d) generally for the better carrying out of the provisions of this Act.

Prosecutions of Political Refugees under Immigration Law to Require Consent of the Attorney-General.[8]

15.   No prosecution for a contravention of the Immigration Law shall be instituted or continued against –

    (a)   an immigrant who is summoned to appear before a committee under the provisions of section 5 (1) (a), pending the decision of the Minister under section 8 (1); or

    (b)   a recognized refugee;

without the written consent of the Attorney-General, and in giving or withholding such consent the Attorney-General shall have regard to the provisions of the Convention.

SCHEDULE[9]

(Section 2)

Definition of the term 'Political refugee'

1.   Subject to the provision of this Schedule 'political refugee' means a person who, owing to well-founded fear of being persecuted for reasons of race, religion, nationality, membership of a particular social group or political opinion, is outside the country of his nationality and is unable or, owing to such fear, is unwilling to avail himself of the protection of that country; or who, not having a nationality and being outside the country of his former habitual residence is unable or, owing to such fear, is unwilling to return to it.

2.   In the case of a person who has more than one nationality, the term 'the country of his nationality' shall mean each of the countries of which he is a national, and a person shall not be deemed to be lacking the protection of the country of his nationality if, without any valid reason based on well-founded fear he has not availed himself of the protection of one of the countries of which he is a national.

Bill No. 71 of 1970

THE REFUGEES (RECOGNITION AND CONTROL) (AMENDMENT)
BILL, 1970

(Published on the 20th November, 1970)

MEMORANDUM

A draft of the above Bill which it is proposed to present to the National Assembly is set out below.

2.   Clause 2 of the Bill applies the provisions of the Employment of Visitors Act, 1968, to refugees. It is, however, provided that the restriction whereby

visitors may not hold work permits for more than 12 months shall not apply to refugees and the Minister may instruct that in the appropriate cases refugees be issued with work permits for a period in excess of 12 months. If such provision were not made it would mean that refugees who have no other country to which to go would have to remain here possibly indefinitely on public assistance. It is considered that where a refugee has a contribution to offer to Botswana in the form of a special skill such as teaching or nursing he should be allowed to make that contribution provided that in so doing he is not doing a job for which a citizen is available.

3.   Clause 3 of the Bill amends the Employment of Visitors Act, 1968, by removing refugees from the definition of visitor.

E.S. MASEL
Minister of State

A BILL
entitled

AN ACT TO AMEND THE REFUGEES (RECOGNITION AND CONTROL) ACT, 1967, BY MAKING PROVISION AS TO THE CONTROL OF EMPLOYMENT OF REFUGEES.

Date of Assent:
Date of Commencement:

ENACTED by the Parliament of Botswana.

Citation
1.   This Act may be cited as the Refugee (Recognition and Control) (Amendment) Act, 1970.

Insertion of new section in Act 8 of 1967
2.   The Refugees (Recognition and Control) Act of 1967, is amended by the insertion therein immediately after section 13 thereof of a new section as follows –

18. A (1)   Subject to the provisions of subsection (2) sections 3 to 6 of the Employment of Visitors Act, 1968, shall apply to refugees as they apply to visitors and any regulations made under the provisions of

section 7 of that Act shall, unless the context otherwise requires, apply to refugees as they apply to visitors.

(2) Notwithstanding anything contained in subsection (1) the Minister may, in his discretion, instruct that a work permit issued to a refugee shall be renewed for such period or periods as he may deem fit notwithstanding that any such renewal will have the effect of extending the validity of such permit for more than twelve months.

Amendment of Act 19 of 1968

3.   Section 2 of the Employment of Visitors Act, 1968, is amended by the insertion immediately after the words 'as amended' appearing in the definition of visitors therein of the words 'but shall not include a refugee'.

Notes

1   Definition inserted by Act No. 37 of 1967.
2   Definition subsituted by Act No. 37 of 1967.
3   Section subsituted by Act No. 37 of 1967.
4   Proviso inserted by Act No. 37 of 1967.
5   Subsection amended by Act No. 37 of 1967.
6   Amended by Act No. 37 of 1967.
7   Ibid.
8   Inserted by Act No. 37 of 1967.
9   Inserted by Act No. 37 of 1967.

Appendix 7: The Refugees (Control) Act, 1970 of the Republic of Zambia

GOVERNMENT OF ZAMBIA ACT
No. 40 of 1970

Date of Assent: 28th August, 1970.
An Act to make provision for the control of refugees; and for matters incidental thereto

4th September 1970

Enactment.
ENACTED by the Parliament of Zambia.

Short Title.
1.   This Act may be cited as the Refugee (Control) Act, 1970.

Interpretation.
2. In this Act, unless the context otherwise requires –
    'authorized' means authorized by the Minister;
    'authorized officer' means the Commissioner and includes a refugee officer, an authorized police officer, an authorized officer of the Zambia Prison Service or of the Defence Force, and any public officer for the time being designated by the Minister as an authorized officer;
    'Commissioner' means the person for the time being holding or acting in the public office of Commissioner for Refugees;
    'Minister' means the member of the Cabinet responsible for matters relating to refugees;
    'Prescribed' means prescribed by the Minister;

'reception area' means an area declared as such under section four;
'refugee' means a person belonging to a class of persons to whom a declaration under section three applies;
'refugee officer' means any person for the time being holding or acting in the public office of refugee officer;
'refugee settlement' means a refugee settlement established under section four.

Declaration of refugees.
3. (1) Subject to the provisions of subsection (2), the Minister may declare, by order, any class of persons who are or prior to their entry into Zambia were, ordinarily resident outside Zambia to be refugees for the purposes of this Act.

(2) A declaration under subsection (1) shall not apply to –

(a) a citizen of Zambia;

(b) any person entitled in Zambia to diplomatic immunity;

(c) any person in the employment of any state, government or local authority outside Zambia, or of any organization to which section four of the Diplomatic Immunities and Privileges Act, 1965, applies, who enters Zambia in the course of his duties;

(d) any member of a class of persons declared by the Minister, by order, not to be refugees for the purposes of this Act.

(3) If any question arises in any proceedings, or with reference to anything done or proposed to be done, under this Act as to whether any person is a refugee or not, or is a refugee of a particular category or not, the onus of proving that such person is not a refugee or, as the case may be, is not a refugee of a particular category, shall lie upon that person.

Reception areas and Refugee Settlements
4. (1) The Minister may declare any part of Zambia to be an area for the reception or residence of any refugees or category thereof.

(2) The Minister may establish in any reception area a refugee settlement for refugees of any category thereof, and may appoint a refugee officer to be in charge of such settlement.

Places of entry or Departure.
5. (1) The Minister may, by order in writing –

(a) direct that any refugee entering or leaving Zambia shall enter or leave by specified routes or at specified places;

(b) direct that any refugee moving from one part of Zambia to another shall move by specified routes.

(2) Orders made under this section may be subject to such terms and conditions as the Minister may think fit.

(3) Any refugee who contravenes an order under this section or the terms or conditions thereof shall be guilty of an offence against this Act.

Registration of refugees and identity cards.

6. (1) Every refugee shall, within such period as may be prescribed, present himself for registration under this section in such manner and to such authority as may be prescribed.

(2) Every refugee shall, upon registration under this section, be issued with an identity card in such form and containing such particulars as may be prescribed, and shall keep such identity card in his possession at all times which in Zambia.

(3) The Minister shall cause registers to be kept for the purposes of this section in such form as the Minister may from time to time determine.

(4) Any refugee who contravenes any of the provisions of this section shall be guilty of an offence against the Act.

Restriction on the possession of firearms, weapons, etc, by refugees.

7. (1) No refugee shall, while in Zambia, acquire or be in possession of any firearm or ammunition.

(2) A refugee who brings any firearm or ammunition into Zambia shall immediately surrender such firearm or ammunition to an authorized officer.

(3) The Minister may, by order in writing, direct that any refugee shall, within such time as may be specified in the order, surrender to an authorized officer any weapon, or any instrument or tool so specified which is capable of being used as a weapon and which is in or comes into his possession, unless the possessor thereof has written authority to retain the same signed by an authorized officer.

(4) Any refugee who –

(a) contravenes the provisions of subsection (1);

(b) fails to surrender any firearm, ammunition, weapon, instrument or tool in accordance with this section or any other made hereunder;

shall be guilty of an offence and shall be liable on conviction to imprisonment for a period not exceeding two years.

(5) In this section, 'firearm' and 'ammunition' have the meanings respectively assigned thereto in the Firearms Act, 1969, but save as aforesaid nothing in that Act shall apply in relation to a refugee.

Detention and slaughter of animals.
8. (1) The Minister may, by order, direct that any animal imported from outside Zambia by any refugee shall be kept in such place as he may direct or, if it appears to the Minister necessary or expedient in the interests of the health of persons or animals, that any such imported animal shall be slaughtered or otherwise disposed of.

(2) Where an animal imported by a refugee is sold, or is slaughtered and the meat or carcass thereof sold, in pursuance of a direction under subsection (1), the proceeds of the sale less the expenses of the sale shall be paid to such refugee; Provided that where in any case it is not reasonably practicable to make payment as aforesaid such proceeds hall be paid into a fund which shall be used for the benefit of refugees.

(3) Any person who obstructs the carrying out of any direction given under this section shall be guilty of an offence against this Act.

Use of Vehicles.
9. (1) No vehicle in which a refugee enters Zambia, or which is acquired by or comes into the possession of a refugee while in Zambia, shall be used in Zambia by such refugee save with the permission in writing of an authorized officer or otherwise than in accordance with the terms of such permission.

(2) It shall be a condition for the granting of permission under this section for the use of a vehicle that an authorized officer may, for so long as may be necessary for the purpose of providing transport for refugees or of moving any stores or equipment for the use of refugees, take possession of such vehicle at any time and may authorize its use by any person for any of the said purposes.

Deportation of Refugees.
10.(1) The Minister may at any time order any refugee to return by such means or route as he shall direct to the territory from which he entered Zambia.

(2) A court convicting any refugee of an offence under the provisions of this section may order the deportation of such refugee to the territory from which he entered Zambia.

(3) Where any person is ordered to return to the territory from which he entered Zambia or to be deported under subsection (1) or (2) he may be held in custody and deported in accordance with such order.

(4) No order shall be made under subsection (1) or (2) in respect of a refugee if the Minister or the court, as the case may be, is of the opinion that such refugee may be tried, or detained or restricted or punished without trial for an offence of a political character after arrival in the territory from which he came or is likely to be the subject of physical attack in such territory.

(5) Any refugee who fails to comply with an order made under subsection (1) shall be guilty of an offence against this Act.

(6) Where an order is made under this section in respect of a refugee who has been present in Zambia for a continuous period of not less than three months immediately prior to the making of the order, the authority making the order shall inform the refugee, or cause him to be informed, that he may make representations against his deportation on the grounds that he is in danger of being tried, or detained or restricted or punished without trial, for an offence of a political character after arrival in the territory from which he came or is in danger of physical attack in such territory. A refugee to whom this subsection applies who wishes to make such representations shall make them forthwith to the person by whom he is so informed and that person shall reduce such representations to writing and forward them to the Minister; and the Minister shall consider the same and determine whether or not the refugee shall be deported in accordance with the order in that behalf or whether that order shall be revoked, and where the Minister determines that the order shall be revoked, he shall have the power to revoke the same. Pending the determination of the Minister on any such representations, the order for the deportation of the refugee shall be suspended.

Permits to remain in Zambia.

11. (1) No refugee shall remain in Zambia –
   (a) unless within seven days of his entering Zambia he is issued with a permit to remain by an authorized officer;
   (b) unless he complies with the terms or conditions from time to time annexed to such permit by an authorized officer.

(2) An authorized officer shall not refuse a refugee a permit under this section if the officer has reason to believe that the refusal of a permit will necessitate the return of the refugee to the territory from which he entered

Zambia and that the refugee may be tried, or detained or restricted or punished without trial, for an offense of a political character after arrival in that territory or is likely to be the subject of physical attack in that territory; but, save as aforesaid, such authorized officer may in his discretion and without assigning any reason refuse to issue a permit.

(3) If a refugee fails to obtain or is refused a permit in accordance with this section, his present in Zambia shall be unlawful.

Requirement to reside in reception area or refugee settlement.

12. (1) The Minister may –
   (a) by order, require any refugee to reside within a reception area or refugee settlement;
   (b) require any refugee who is within a reception area or refugee settlement to remove to and reside in some other place being a reception area or refugee settlement.

(2) Any refugee to whom an order made under this section applies who –
   (a) fails to take steps forthwith to comply with such order; or
   (b) fails to move to or take up residence in a reception area or refugee settlement in accordance with such order with reasonable dispatch; or
   (c) having arrived at a reception area or a refugee settlement in pursuance of such order, leaves or attempts to leave such area or settlement except in pursuance of some other order made under this section;

shall, unless he is in possession of a permit issued in that behalf under subsection (3), be guilty of an offence against this Act.

(3) An authorized officer may issue a permit to any refugee to whom an order made under subsection (1) applies, authorizing him –
   (a) to reside in a reception area elsewhere than in the refugee settlement to which such order refers;
   (b) to leave a reception area in which he has been required to reside.

(4) An authorized officer may issue a permit under this section subject to such terms and conditions as he thinks fit, and, without prejudice to the generality of the foregoing, he may, where he issues a permit under paragraph (b) of subsection (3), specify the destination to which such refugee may proceed.

(5) Any refugee to whom a permit has been issued under this section who fails to comply with the terms and conditions thereof shall be guilty of an offence against this Act.

Control of refugee settlements.

13.(1) The Minister may make rules, and the Commissioner may issue directions not inconsistent with such rules, for the control of refugee settlements and, without prejudice to the generality of the foregoing, such rules and directions may make provision in respect of all or any of the following matters;

(a) the organization, safety, discipline and administration of such settlements;

(b) the reception, treatment, health and well-being of refugees;

(c) the powers of refugee officers in respect of such settlements.

(2) A refugee officer may give such orders or directions, either orally or in writing, to any refugee as may be necessary or expedient for the following purposes, that is to say –

(a) to ensure that any refugee settlement is administered in an orderly and efficient manner;

(b) to ensure the performance of any work or duty necessary for the maintenance of essential services in any refugee settlement or for the general welfare of the refugees therein;

(c) to ensure that all proper precautions are taken to preserve the health and well-being of the refugees therein;

(d) to preserve orderly conduct and discipline in any refugee settlement.

(3) Any refugee who –

(a) without a permit in that behalf issued under section twelve, leaves or attempts to leave a refugee settlement in which he has been order to reside; or

(b) in a refugee settlement disobeys any rules made by the Minister, any direction of the Commissioner or any order or direction of a refugee officer, made or given under this section; or

(c) in a refugee settlement conducts himself in a manner prejudicial to good order and discipline;

shall be guilty of an offence against this Act.

Restrictions relating to refugee settlements.

14.(1) No person other than a refugee required to reside or residing in, or a person employed in, a refugee settlement, shall enter or be within such settlement except with the general or special permission of the Minister, the Commissioner or a refugee officer.

(2) No person other than the Commissioner or a refugee officer may in a refugee settlement address an assembly or meeting of more than ten refugees whether or not such meeting is held in a public place.

(3) Any person who contravenes the provisions of this section shall be guilty of an offence against this Act.

Offences and Penalties.

15.(1) Any refugee who fails to obey any lawful order of the Commissioner or a refugee officer or who obstructs an authorized officer in the exercise of his powers under this Act shall be guilty of an offence and shall be liable to conviction to imprisonment for a period not exceeding three months.

(2) Any person who is guilty of an offence against this Act for which no penalty is specifically provided shall be liable on conviction to imprisonment for a period not exceeding three months.

Arrest and use of force.

16.(1) An authorized officer may arrest without warrant any refugee reasonably suspected by the authorized officer of having committed or attempted to commit an offence against this Act.

(2) An authorized officer and any person acting with the authority of an authorized officer may use such force, including the use of firearms, as may be reasonably necessary to compel any refugee to comply with any order or direction made or given under this Act in relation to such refugee.

Protection of Officers.

17. No act or thing done or omitted to be done by any authorized officer or other person shall, if the act or omission was done or omitted bonafide while acting in the execution of his duty under this Act, subject him personally to any liability, action, claim or demand whatsoever.

Manner of publishing rules, order, etc.

18.(1) Save as provided in this section or any other law, rules, orders or directions under this Act may be published in such manner as the authority making the same considers appropriate in order to bring the same to the notice of the persons to whom they apply or who are affected thereby, and shall not be required to be published in the Gazette.

(2) Any order or direction made or given under this Act which is applied or directed to a particular person shall, if in writing, be served on, or if not in writing, be given to, that person personally.

(3) Every declaration under section three shall be published in the Gazette.

Appendix 8: List of States Party to the UN Convention 1951 and/or the Protocol 1967 Relating to the Status of Refugees as at July 1987

States party to the 1951 UN Convention	99
States party to the 1967 Protocol	100
States party to both the 1951 Convention and the 1967 Protocol	96
States party to either one or both of these instruments	103

Africa

Algeria
Angola

Benin
Botswana
Burkina Faso
Burundi

Cameroon
Cape-Verde (P)
Central African Republic
Chad
Congo
Cote d'Ivoire

Djibouti

Egypt
Equatorial Guinea
Ethipia

Gabon
Gambia
Ghana
Guinea
Guinea-Bissau

Kenya

Lesotho
Liberia
Madagascar (C)*
Mali
Morocco
Mozambique (C)

Africa (cont'd)

Niger
Nigeria

Rwanda

Sao Tome and Principe
Senegal
Seychelles
Sierra Leone
Somalia

Sudan
Swaziland (P)

Togo
Tunisia

Uganda
United Republic of Tanzania

Zaire
Zambia
Zimbabwe

Americas

Argentina

Bolivia
Brazil*

Canada
Chile
Colombia
Costa Rica

Dominican Republic

Ecuador
El Salvador

Guatemala

Haiti

Jamaica

Nicaragua

Panama
Paraguay*
Peru

Surinam

United States of America (P)

Uruguay

Venzuela (P)

Asia

China

Iran (Islamic Republic of)
Israel

Japan

Phillipines

Yemen

Europe

Austria

Belgium

Cyprus

Denmark[1]

Finland
France[2]

Germany, Federal Republic of[3]
Greece

Holy See

Iceland
Ireland
Italy*

Leichtenstein

Malta*
Monaco (C)*
Netherlands[4]
Norway

Portugal

Spain
Sweden
Switzerland

Turkey*

United Kingdom[5]

Yugoslavia

Oceania

Australia[6]

Fiji

New Zealand

Papua New Guinea

Tuvalu

Notes

*   The seven states marked with an asterisk – Brazil, Italy, Madagascar, Malta, Monaco, Paraguay and Turkey – have made a declaration in accordance with Article 1 (B) 1 of the 1951 Convention to the effect that the words 'events occurring before 1 January 1951' in Article 1, Section A, should be understood to mean 'events occurring *in Europe* before 1 January 1951'. All other party states apply the Convention without geographical limitation. The following five states have expressly maintained their declarations of geographical limitation with regard to the 1951 Convention upon acceding to the 1967 Protocol: Brazil, Italy, Malta, Paraguay and Turkey. Madagascar and Monaco have not yet adhered to the 1967 Protocol.

(C) Party to the 1951 Convention only.

(P) Party to the 1967 Protocol only.

1   Denmark declared that the Convenion was also applicable to Greenland.

2   France declared that the Convention applied to all territories for the international relations of which France was responsible.

3   The Federal Republic of Germany made a separate declaration stating that the Convention and the Protocol also applied to Land Berlin.

4   The Netherlands extended application of the Protocol to Aruba.

5   The United Kingdom extended application of the Convention to the following territories for the conduct of whose international relations the governemnt of the United Kingdom is responsible: Channel Islands, Falkland Islands (Malvinas), Isle of Man, St Helens.

     The United Kingdom declared that its accession to the Protocol did not apply to Jersey, but extended its application to Montserrat.

6   Australia extended application of the Convention to Norfolk Island.

Appendix 9: List of States Party to the OAU Convention 1969 Governing the Specific Aspects of Refugee Problems in Africa (January 1987)

Algeria

Angola

Benin

Burkino Faso

Burundi

Cameroon

Central African Republic

Chad

Congo

Egypt

Equatorial Guinea

Ethiopia

Gabon

Gambia

Ghana

Guinea

Liberia

Libyan Aran Jamahiriya

Mali

Mauritania

Morocco

Niger

Nigeria

Rwanda

Senegal

Seychelles

Somalia

Sudan

Togo

Tunisia

United Republic of Tanzania

Zaire

Zambia

Zimbabwe

Appendix 10: African Independent States

	Year of independence[1]	Population[2] (1978 estimate)	Population growth rate percentage 1988[3]
Algeria	1962	17,825,000	2.0
Angola	1975	6,831,000	2.6
Benin	1960	3,341,000	3.0
Botswana	1966	750,000	3.4
Burundi	1962	4,068,000	3.3
Cameroon	1960	7,981,000	2.6
Cape Verde	1975	311,000	2.5
Central African Republic	1960	1,912,000	2.5
Chad	1960	4,285,000	2.0
Comoros	1975	313,000	3.3
Djibouti	1977	242,000	3.0
Egypt*	1922	38,362,000	2.8
Equatorial Guinea	1968	327,000	1.9
Ethiopia*	–	30,037,000	2.1
Gabon	1960	1,300,000	1.9
Gambia, The	1965	569,000	2.5
Ghana	1957	10,755,000	3.1
Guinea	1958	4,762,000	2.4
Guinea Bissau	1974	949,000	1.0
Ivory Coast (Cote d'Ivoire)	1960	7,205,000	3.6
Kenya	1963	14,658,000	4.1
Lesotho	1966	1,268,000	2.8
Liberia*	1847	1,717,000	3.2
Libya	1951	2,678,000	3.1
Madagascar	1960	8,776,000	3.1
Malawi	1964	5,711,300	3.3
Mali	1960	6,341,000	2.9
Mauritania	1960	1,544,000	2.7

	Year of independence[1]	Population[2] (1978 estimate)	Population growth rate percentage 1988[3]
Morocco	1956	19,168,000	2.6
Mozambique	1975	9,899,000	2.6
Namibia	1990 (1986)	1,180,000	3.1
Niger	1960	4,986,000	2.9
Nigeria	1960	80,862,000	2.9
Rwanda	1960	4,582,000	3.4
Sao Tome and Principe	1975	83,000	2.7
Senegal	1960	5,332,000	2.6
Seychelles	1976	63,000	1.8
Sierra Leone	1961	3,220,000	2.4
Somalia	1960	3,446,000	2.6
Sudan, The	1955	17,141,000	2.8
Swaziland	1968	528,000	3.4
Tanzania (inc. Zanzibar)	1961 (1963)	16,590,000	3.6
Togo	1960	2,449,000	3.4
Tunisia	1956	6,071,000	2.4
Uganda	1962	12,724,000	3.4
Upper Volta (Burkina Faso)	1960	6,464,000	2.8
Zaire	1960	26,478,000	3.1
Zambia	1964	5,514,000	3.7
Zimbabwe	1980	–	3.6

Notes

* Countries that were independent before 1950.

1 *UN Population and Vital Statistics Report*: official country figures.
2 Ibid.
3 *1989 World Population Data Sheet*, Population Reference Bureau Inc., Washington DC.

Appendix 11: Statistical Report on the Movement of Refugees in Africa as at 31 December 1988

Country of asylum	Country of origin		Numbers
Horn of Africa			
Djibouti UNHCR budget for 1989: US$ 1,628,000	Mostly from Ethiopia		7,000
Ethiopia UNHCR budget for 1989: US$ 605,800	Somalia Sudan Various origins	5000,000 324,000 500	824,500
Somalia UNHCR budget for 1989: US$ 34,652,200	Mostly from Ethiopia		840,000
Sudan UNHCR budget for 1989: US$ 38,217,500	Ethiopia Chad Uganda Zaire	630,000 50,000 15,000 5,000	700,000
Southern Africa			
Angola UNHCR budget for 1989: US$ 5,245,000	Namibia Zaire South Africa	74,000 12,200 10,000	96,200

Country of asylum	Country of origin		Numbers
Botswana			3,542
UNHCR budget for 1989:	Zimbabwe	3,000	
US$ 1,292,200	Angola	425	300
	Namibia	117	
Lesotho	South Africa		
UNHCR budget for 1989:			16,800
US$ 460,600			
Swaziland	Mozambique	10,300	
UNHCR budget for 1989:	South Africa	6,500	600,000
US$ 1,696,000			
Malawi	Mostly from Mozambique		
UNHCR budget for 1989:			400
US$ 18,967,000			
Mozambique	South Africa		
UNHCR budget for 1989:	Chile		
US$ 2,850,000	East Timor		146,000
Zambia	Angola	97,000	
UNHCR budget for 1989:	Mozambique	27,000	
US$ 2,992,000	Zaire	9,000	
	Namibia	6,900	
	South Africa	3,200	
	Various origins	2,900	166,200
Zimbabwe	Mozambique	166,000	
UNHCR budget for 1989:	South Africa)		
US$ 5,128,000	Malawi)		
	Namibia)	200	
	Uganda)		

Country of asylum	Country of origin		Numbers
East Africa			11,500
Kenya UNHCR budget for 1989: US$ 2,382,400	Uganda Ethiopia Rwanda Various origins	6,300 2,200 2,000 1,000	266,900
Tanzania UNHCR budget for 1989: US$ 3,950,000	Burundi Mozambique Rwanda Zaire Various origins	154,400 72,000 23,000 16,000 1,500	96,000
Uganda UNHCR budget for 1989: US$ 2,434,000	Rwanda Sudan	84,000 12,000	270,000
Central Africa			
Burundi UNHCR budget for 1989: US$ 498,000	Mostly from Rwanda		53,600
Cameroon UNHCR budget for 1989: US$ 1,665,100	Chad Namibia	53,531 69	4,300
Central African Republic UNHCR budget for 1989: US$ 680,000	Mostly from Chad		2,100
Congo UNHCR budget for 1989: US$ 579,000	Chad Zaire Central African Republic	1,605 336 159	A small number
Gabon UNHCR budget for 1989: US$ 85,000	Various origins		76,500

Country of asylum	Country of origin		Numbers
Rwanda			320,000
UNHCR budget for 1989:			
US$ 926,000	Mostly from Burundi		
Zaire			
UNHCR budget for 1989:	Angola	298,700	
US$ 3,947,00	Burundi and		
	Rwanda	16,000	18,189
	Uganda	4,000	
	Various origins	1,300	

West Africa
UNHCR budget for 1989:
US$ 5,502,800

Benin	-3,000)			
Burkina Faso	266)			
Cote d'Ivoire	-783)	Chad	7,600	
Ghana	-156)	Ghana	3,800	
Liberia	-166)	Guinea-Bissau	5,000	
Nigeria	-5,085)	Various origins	1,789	4,104
Senegal	-101)			
Togo	-3,447)			

North Africa
UNHCR budget for 1989:
US$ 5,139,300

Algeria	)	Western Sahara
Egypt	)	Middle East
Morocco	)	Europe
Tunisia	)	Ethiopia
		Sudan
		Namibia
		Chad

Note

The numbers of refugees are provided by governments based on their own records and methods of estimation.

Source: *Refugees*, December 1988, Special Issue.

Appendix 12: Africa – Four Million Refugees

Mediterranean Sea

Tunisia

Morocco

Algeria (165,000)

Libyan Arab Jamahiriya

Egypt (1,100)

Red Sea

Senegal

Mauritania

Mali

Niger

Chad

Sudan (974,200)

Gambia

Guinea

Burkina Faso

Nigeria (4,000)

Central African Republic (7,610)

Ethiopia (164,000)

Guinea Bissau

Cote d'Ivoire

Somalia (840,000)

Liberia

Ghana

Benin (4,040)

Congo 1,990

Kenya (8,045)

Sierra Leone

Togo (3,500)

Cameroon (7,350)

Gabon

Uganda (112,270)

Equatorial Guinea

Zaire (301,000)

Tanzania (United Rep. of) (220,265)

Rwanda (19,380)

Burundi (198,000)

Angola (82,180)

Zambia (133,600)

Malawi (210,000)

Madagascar

Atlantic Ocean

Namibia

Zimbabwe (65,000)

Mozambique

Botswana (5,180)

Swaziland (12,650)

South Africa (150,000)

Lesotho (4,000)

Indian Ocean

Appendix 13: Convention Travel Document for Refugees (Specimen)

ANNEX

Special Travel Document

The document will be in booklet form (approximately 15 x 10 centimetres).

It is recommended that it be so printed that any erasure or alteration by chemical or other means can be readily detected, and that the words 'Convention of 28 July 1951' be printed in continuous repetition on each page, in the language of the issuing country.

(Cover of booklet)

TRAVEL DOCUMENT
(Convention of 28 July 1951)

No. _____

(1)
TRAVEL DOCUMENT
(Convention of 28 July 1951)

This document expires on _____
unless its validity is extended on renewed.

Name _____

Forename(s) _____

Accompanied by _____ child (children)

1.   This document is issued solely with a view to providing the holder with a travel document which can service in lieu of a national passport. It is without prejudice to and in no way affects the holder's nationality.

2.   The holder is authorized to return to _____
_____ [state here the country whose authorities are issuing the document] on or before _____ unless some later date is hereafter specified.

[The period during which the holder is allowed to return must not be less than three months]

3.   Should the holder take up residence in a country other than that which issued the present document, he must, if he wishes to travel again, apply to the competent authorities of his country of residence for a new document. [The old travel document shall be withdrawn by the authority issuing the new document and returned to the authority which issued it.][1]

(This document contains —————— pages, exclusive of cover.)

_____

[1]   The sentence in brackets to be inserted by Governments which so desire.

(2)

Place and date of birth —————————————————————

Occupation ————————————————————————

Present residence ——————————————————————

*Maiden name and forename(s) of wife ————————————

*Name and forename(s) of husband ————————————

Description

Height   ——————————————————

Hair   ——————————————————

Colour of Eyes   ——————————————————

Nose   ——————————————————

Shape of face   ——————————————————

Complexion ——————————————————

Special peculiarities   ——————————————————

Children accompanying holder

Name	Forename(s)	Place and date of birth	Sex
——————	——————	——————	————
——————	——————	——————	————
——————	——————	——————	————
——————	——————	——————	————

* Strike out whichever does not apply

(This document contains —————— pages, exclusive of cover.)

(3)

Photograph of holder and stamp of issuing authority
Finger-prints of holder (if required)

Signature of holder ——————————————————————

(This document contains —————— pages, exclusive of cover.)

(4)

1.   This document is valid for the following countries:

_____
_____
_____
_____

2.   Document or documents on the basis of which the present document is issued:

_____
_____
_____

Issued at ─────────────────
Date ─────────────────────

                         Signature and stamp of authority
                              issuing the document:

Fee paid:
         (This document contains ──────── pages, exclusive of cover.)

(5)
Extension or renewal of validity

Fee paid: ───────────── From ──────────────────
                        To   ──────────────────
Done at ─────────────── Date ──────────────────

                         Signature and stamp of authority
                         extending or renewing the validity
                              of the document:

Extension or renewal of validity

Fee paid: ───────────── From ──────────────────
                        To   ──────────────────
Done at ─────────────── Date ──────────────────

                         Signature and stamp of authority
                         extending or renewing the validity
                              of the document:

         (This document contains ────────pages, exclusive of cover.)

(6)
Extension or renewal of validity

Fee paid: ───────────── From ──────────────────
                        To   ──────────────────
Done at ─────────────── Date ──────────────────

Signature and stamp of authority
extending or renewing the validity
of the document:

Extension or renewal of validity

Fee paid: _____   From _____

                                  To   _____

Done at _____   Date _____

Signature and stamp of authority
extending or renewing the validity
of the document:

(This document contains _____ pages, exclusive of cover.)

(7–32)

Visas

The name of the holder of the document must be repeated in each visa.
(This document contains _____ pages, exclusive of cover.)

SCHEDULE

Paragraph 1

1.   The travel document referred to in Article 28 of this Convention shall
be similar to the specimen annexed hereto.
2.   The document shall be made out in at least two languages, one of which
shall be English or French.

Paragraph 2

Subject to the regulations obtaining in the country of issue, children may be
included in the travel document of a parent or, in exceptional circumstances,
of another adult refugee.

Paragraph 3

The fees charged for issue of the document shall not exceed the lowest scale
of charges for national passports.

Paragraph 4

Save in special or exceptional cases, the document shall be made valid for the largest possible number of countries.

Paragraph 5

The document shall have a validity of either one or two years, at the discretion of the issuing authority.

Paragraph 6

1.   The renewal or extension of the validity of the document is a matter for the authority which issued it, so long as the holder has not established lawful residence in another territory and resides lawfully in the territory of the said authority. The issue of a new document is, under the same conditions, a matter for the authority which issued the former document.

2.   Diplomatic or consular authorities, specially authorized for the purpose, shall be empowered to extend, for a period not exceeding six months, the validity of travel documents issued by their Governments.

3.   The Contracting States shall give sympathetic consideration to renewing or extending the validity of travel documents or issuing new documents to refugees no longer lawfully resident in their territory who are unable to obtain a travel document from the country of their lawful residence.

Paragraph 7

The Contracting States shall recognize the validity of the documents issued in accordance with the provisions of Article 28 of this Convention.

Paragraph 8

The competent authorities of the country to which the refugee desires to proceed shall, if they are prepared to admit him and if a visa is required, affix a visa on the document of which he is the holder.

Paragraph 9

1.   The Contracting States undertake to issue transit visas to refugees who have obtained visas for a territory of final destination.
2.   The issue of such visas may be refused on grounds which would justify refusal of a visa to any alien.

Paragraph 10

The fees for the issue of exit, entry or transit visas shall not exceed the lowest scale of charges for visas on foreign passports.

Paragraph 11

When a refugee has lawfully taken up residence in the territory of another Contracting State, the responsibility for the issue of a new document, under the terms and conditions of Article 28, shall be that of the competent authority of that territory, to which the refugee shall be entitled to apply.

Paragraph 12

The authority issuing a new document shall withdraw the old document and shall return it to the country of issue, if it is stated in the document that it should be so returned; otherwise it shall withdraw and cancel the document.

Paragraph 13

1.   Each Contracting State undertakes that the holder of a travel document issued by it in accordance with Article 28 of this Convention shall be re-admitted to its territory at any time during the period of its validity.
2.   Subject to the provisions of the preceding sub-paragraph, a Contracting State may require the holder of the document to comply with such formalities as may be prescribed in regard to exit from or return to its territory.
3.   The Contracting States reserve the right, in exceptional cases, or in cases where the refugee's stay is authorized for a specific period, when issuing the document, to limit the period during which the refugee may return to a period of not less than three months.

Paragraph 14

Subject only to the terms of paragraph 13, the provisions of this Schedule in no way affect the laws and regulations governing the conditions of admission to, transit through and establishment in, and departure from, the territories of the Contracting States.

Paragraph 15

Neither the issue of the document nor the entries made thereon determine or affect the status of the holder, particularly as regards nationality.

Paragraph 16

The issue of the document does not in any way entitle the holder to the protection of the diplomatic or consular authorities of the country of issue, and does not confer on these authorities a right of protection.

Appendix 14: Distribution of Refugees in Countries of Asylum

	1981[1]	1988[2]	Mainly from:
Somalia	1,500,000	840,000	Ethiopia
Sudan	500,000	807,000	Ethiopia and Uganda
Malawi	–	401,600	Mozambique
Zaire	400,000	320,000	Angola and Sudan
Ethiopia	11,000	310,000	Sudan and Somalia
Burundi	200,000	267,500	Rwanda
Tanzania	150,000	226,200	South Africa and Zaire
Zambia	36,000	146,100	Angola and Namibia
Zimbabwe	–	123,600	Mozambique
Angola	73,000	91,200	Zaire, South Africa and Namibia
Uganda	113,000	87,800	Sudan, Rwanda and Zaire
Cameroon	266,000	53,600	Zaire
Rwanda	–	19,500	Burundi
Djibouti	42,000	13,100	Ethiopia

Notes

1 United Nations, *International Conference on Assistance to Refugees in Africa*, 9–10 April 1981, Geneva.
2 United Nations, High Commissioner for Refugees population map, 1 January 1988. The figures are provided mostly by governments based on their own records and methods of estimation.

Appendix 15: Ratio of Refugees to Local Population in 12 African Countries

Country	Population[i] (millions)	Refugees[ii] (thousands)	Ratio of refugees to local population		
Somalia	4.6	700.0	1	in	7
Djibouti	0.5	31.6	1	in	16
Burundi	4.4	214.0	1	in	21
Sudan	19.9	627.0	1	in	32
Angola	6.8	93.6	1	in	73
Zaire	30.3	325.0	1	in	93
Swaziland	0.6	5.8	1	in	103
Zambia	6.0	58.3	1	in	103
Tanzania	19.9	174.0	1	in	114
Algeria	20.1	167.0	1	in	120
Uganda	13.7	113.0	1	in	121
Lesotho	1.4	11.5	1	in	122

Notes: often the ratio of refugees to local people is even higher than indicated because the refugees are concentrated in particular areas.

Sources: i) Population Reference Bureau Inc., 1982.
ii) UNHCR World Refugee Survey 1982.